BLACK SOFTWARE

BLACK
SOFTWARE

THE INTERNET AND RACIAL JUSTICE, FROM THE AFRONET TO BLACK LIVES MATTER

★

CHARLTON D. MCILWAIN

OXFORD
UNIVERSITY PRESS

OXFORD
UNIVERSITY PRESS

Oxford University Press is a department of the University of Oxford.
It furthers the University's objective of excellence in research, scholarship,
and education by publishing worldwide. Oxford is a registered trade mark of
Oxford University Press in the UK and certain other countries.

Published in the United States of America by Oxford University Press
198 Madison Avenue, New York, NY 10016, United States of America.

© Oxford University Press 2020

Library of Congress Cataloging-in-Publication Data

Names: McIlwain, Charlton D., 1971– author.
Title: Black software : the Internet and racial justice, from the AfroNet
to Black Lives Matter / Charlton McIlwain.
Description: New York, NY : Oxford University Press 2020. | Includes
bibliographical references and index. | Summary: "Black Software, for
the first time, chronicles the long relationship between African
Americans, computing technology, and the Internet. Through new archival
sources and the voices of many of those who lived and made this history,
this book centralizes African Americans' role in the Internet's creation
and evolution, illuminating both the limits and possibilities for using
digital technology to push for racial justice in the United States and
across the globe"—Provided by publisher.
Identifiers: LCCN 2019006334 (print) | LCCN 2019980137 (ebook) |
ISBN 9780190863845 (hardcover ; alk. paper) | ISBN 9780190863852 (pdf) |
ISBN 9780190863869 (epub)
Subjects: LCSH: African Americans—Communication. | African Americans
and mass media. | African Americans—Politics and government—21st century. |
Internet—Political aspects—United States. | Social justice—United States. |
Racism—United States.
Classification: LCC P94.5.A37 M35 2020 (print) | LCC P94.5.A37 (ebook) |
DDC 302.23089/96073—dc23
LC record available at https://lccn.loc.gov/2019006334
LC ebook record available at https://lccn.loc.gov/2019980137

1 3 5 7 9 8 6 4 2
Printed by Sheridan Books, Inc., United States of America

*For all the women and men whose contributions to computing and the
Internet have scarcely been recognized
For all those who labor in the struggle for racial justice*

CONTENTS

★

BOOK TWO

ACKNOWLEDGMENTS

★

This book is a collective project, co-produced with many people that I must acknowledge and thank, despite knowing that acknowledgments and thanks are not nearly enough.

My son, Marcus, and my wife, Raechel, graciously lent me so much time that rightfully belonged to them. Marcus was honest enough to tell me that my first draft of the book was...boring! And Raechel's more sophisticated editorial advice pushed me to write something that was not too sophisticated to be understood by those for whom I wrote this book. I could not have written this book without the love, support, and patience you both showed.

Diana Kamin, Rachel Kuo, and Tristan Beach: I am grateful and indebted to each of you for your interest in and dedication to this project. Your research acumen and your contribution of time ran so far beyond what I paid you for your labor. This book literally would not have been possible without you. Your hard work is etched in every page. Thank you.

I must especially thank the women and men who so generously lent me their time, their words, and their voices—most personally, a few, posthumously. Kamal Al-Mansour, Lee Bailey, Anita Brown, Derrick Brown, Charlene Caruthers, Allen Frimpong, Malcolm Cassell, Farai Chideya, Barry Cooper, E. David Ellington, Tyronne Foy, Ken Granderson, Dream Hampton, Timothy L. Jenkins, William Murrell, Ken Onwere, and Idette Vaughn, I hope I did your stories justice.

I am also indebted to my colleagues and friends whose conversations, ideas, and writings accompanied me as I wrote the book. To many of my colleagues at the Center for Critical Race and Digital Studies—André

Brock, Safiya Noble, Sarah Jackson, Lisa Nakamura, Wendy Chun, Paula Chakravartty, Meredith Broussard, Aymar Jean Christian, Meredith Clark, Deen Freelon, Stephanie Dinkins, Margaret Hu, Jenny Korn, Lori Kiddo Lopez, Shaka McGlotten, Kelli Moore, Anika Navaroli, Alondra Nelson, Desmond Patton, Minh-Ha T. Pham, Kiran Samuel, Catherine Knight-Steele, Robin Stevens, Tonia Sutherland, Brendesha Tynes, Farida Vis, Myra Washington, Anne Washington, and Jessica Eaglin.

Thanks to many other colleagues whose conversations and work help to shape my work. These include Andrea Dixon, Robin Coleman, Aswin Punathambekar, Siva Vaidhyanathan, Rayvon Fouché, Pamela Newkirk, Phil Dalton, David Worth, Nick Couldry, Adrienne Russell, Stephanie Richter Schulte, Matt Powers, Rod Benson, Bilge Yesil, Jessie Daniels, Nabil Echchaibi, Mako Hill, Brett Gary, Arun Kundnani, Lynne Clark, David Goldberg, Finn Brunton, Todd Gitlin, Christina Dunbar-Hester, Gina Neff, Lance Bennett, Carmen Gonzalez, Katy Pearce, Phil Howard.

Special thanks to journalist Emily Parker, who worked on key interviews with and conversations about Black Lives Matter organizational leadership.

Special thanks to my 2015–2016 NYU Humanities Fellows colleagues who provided an engaging audience for the book as it first developed. They include Luke Stark, Narges Bajoghli, Patrick Deer, Valeria G. Castelli, Hi'ilei Julia Hobart, Sean Larson, Anna McCarthy, Andrew Needham, Sean Nesselrode, Lana Povitz, Myisha Priest, Gwynneth C. Malin, Jane Tylus, and Deborah Willis.

Thanks to Katy Boss, Scott Collard and April Hathcock, at NYU Libraries, who helped in many ways with my research for and production of the book. And to archivists at the Moorland-Spingarn Research Center archives at Howard University who so respectfully assisted my search through what seemed like endless boxes of paper. Thanks to Kreg V. Purcell at the National Criminal Justice Reference Service for help locating crucial historical documents. And thanks to Amy Bradley and Max Campbell at IBM Corporate Archives for assistance with securing permission to use IBM images.

I also want to thank danah boyd, Alice Marwick, and other folks affiliated with the Data & Society Research Institute. Serving as an advisor has given me an opportunity to interact with many scholars whose work helped to invigorate my thinking, and the trajectory of this book.

Thanks to Ken Adams who provided comments on a draft of the manuscript.

Finally, much gratitude goes to my editor at Oxford University Press, Angela Chnapko, and publisher Niko Pfund. They believed in this book when it was just a few pages of ideas and a tiny kernel of the story that became *Black Software*.

BLACK SOFTWARE

INTRODUCTION

I returned refreshed from my long, Independence Day weekend, determined to start writing this book. That was Tuesday, July 5, 2016. I'd cleared a couple of hours in the late afternoon, hoping to produce at least a good page or two. I briefly checked Twitter that morning, on the way to my office. Alton Sterling's name repeated as I scrolled through my feed. In the early morning hours, white police officers in Baton Rouge, Louisiana, had tased, subdued, and then shot Sterling six times at point-blank range. The reports that echoed throughout my curated corner on the social media platform said he was unarmed.

I did not write any pages that day. It wasn't that I was distraught by the news. Too much other work had simply stormed its way onto my agenda. And, sad to say, police shootings of unarmed black people in the United States had become all too common to be that emotionally debilitating. Just add Sterling's name to the list, and wait for the inevitable. This time, the inevitable took place the next day. On Wednesday, July 6, police officers in small-town Falcon Heights, Minnesota, stole the life of another black man. Philando Castile was thirty-two years old. Like Sterling the day before, and so many before, Castile posed no threat to law enforcement. They killed him nonetheless.

Four days later, I finally resolved to put ink to paper (as it were), even if just a line or two. I dropped my seven-year-old son off at his dojo. Then

I walked to the Starbuck's coffee shop just outside Barclays Center in Brooklyn. I arrived ready to enjoy, and take advantage of, that kind of Sunday-morning quiet that can help you feel delightfully isolated in a city of millions. I wrested my laptop from its case. But before I could even open a new document, people began to steadily stream into the area. I had come to write. But I found myself in the middle of a Black Lives Matter rally. Throngs of people—infants, toddlers, teenagers, and parents—filled the square. They carried signs saying things like "When will Black lives REALLY matter?" Drivers laid on their horns, piercing my comfortable silence as they crept through the borough's main artery on Atlantic Avenue. Police sirens soon added to the cacophony.

I could not hear the folks leading the gathering, though they spoke valiantly on two not-quite-adequate megaphones. But I knew why they were there. I knew what they wanted. For a moment, my eyes fixated on a black woman walking through the crowd. Her face was pained by much more than an early Sunday morning rise from bed. She carried a sign whose writing was obscured by a larger, framed photograph. She carried it close to her, clutched in one hand at the bottom. The sign did not identify the framed young man by name. No doubt, though, he was one of the hundreds of black victims who had died at the hands of police officers in just the first six months of 2016 alone.

My laptop open, I sat, wishing I could write something about how to change the circumstances that brought this crowd together that day. But better and more courageous folks than I have tried—to no avail—for decades. The crowd soon dispersed, and I began to write, placated by the knowledge that while what I wrote might not change the persistent conditions black people face with the modern criminal justice system, it might help us better understand the latest movement to make that change. Ironically, as I set off to retrieve my son, I spotted a friend, mom to one of my son's classmates. "Oh, were you here for the rally?" she asked. "Not intentionally," I replied, a little happy to see her, a little sad my writing time had ended. When I told her about the book and my grand plans to begin writing it, she started dropping names. "Do you know Farai Chideya?" she asked. "We went to high school together; she was my girl..." "I just talked to her a couple of months ago," I said. "Omar Wasow?" she asked, recalling a 1990s Vibe music festival where he introduced her to a revolutionary new technology that would help her continue her conversation with an acquaintance she'd just met.

The irony wasn't just that I spotted a friend in a crowd of hundreds. It was the fact that this book is both about that rally I stumbled onto, and the names my friend recalled. Decades separate each of them, but their stories are inextricably linked.

In 2015, two colleagues—Deen Freelon and Meredith Clark—and I had set out to better understand how Black Lives Matter emerged. Our report, *Beyond the Hashtags, the Online Struggle for Offline Justice*, crystalized then–NAACP president Cornell Brooks's sentiment: "this isn't your grandparents' civil rights movement." Our study showed us that Ferguson, Missouri birthed Black Lives Matter. It told us that Twitter named Michael Brown for the world. Traditional news media outlets were a day late, and when they did arrive, Twitter was their primary source for information. It afforded a twenty-four-hour glimpse into a radical new way of making news, of telling stories—giving unfiltered voice to those whose voices are traditionally unheard, ignored, or silenced.

Digital media—social networking platforms, blogging platforms, the open Web, user-generated and circulated images—all afforded Ferguson citizens, racial justice activists, organizers, young people, and others a powerful way to counter the Hollywood and newsmaker image machine—one that historically and continuously casts African Americans as criminal, intellectually deficient, and culturally deviant.

Trayvon Martin, Renisha McBride, Eric Garner, Mike Brown, Sandra Bland, and many more women and men like them did not become the usual footnotes in the thick annals of black, violent, drug-ridden thugs the American public would easily dismiss. Legions of citizens armed with cellphone cameras and an Internet connection shoved a steel pipe through the cogs driving the American image machine. They flooded the Web with images that painted the historically most feared—black people, and black men more particularly—as sympathetic victims. Online reporting cast them as victims of police bias, violence, and police militarism run amuck. News media framed black people as targets of a larger system built, and deliberately designed, to prey on and disadvantage poor people of color at almost every turn.

When we look at today's racial justice movement's success consider this. Movement participants built dense *and* diverse networks between affiliated activists, journalists, public officials, and mass groups of anonymous strangers to both hijack and resist media influence and power. They produced, and widely distributed, compelling, consumable, and usable content across digital space to inform and organize mass audiences. They manipulated algorithms to influence search outcomes. They enhanced public visibility and compelled attention to racial justice causes and actions. They used digital hardware and software to obfuscate state surveillance systems and watch the watchers. They created digital stylistics and protected spaces that produced personal and collective pleasure, instigated conversation and debate, and established community. These compose today's racial justice movement's digital toolkit. Mastering these tools has produced the most visible, sustained, and vociferous movement toward racial justice we've witnessed in the United States since the 1960s.

But at the same time, it is both inaccurate and disingenuous to reduce today's movement to those who merely master its digital tools. The people who have widely used and mastered the digital tools that fueled Black Lives Matter and today's broader racial justice movement reflect and required a prior technical and political socialization. If this is true, then where did today's digitally revolutionized racial justice movement come from? Who engineered it? What is its signature and what effect has it had on our society? Where will it take us from here?

I wanted to answer these questions. So I began to search. I spoke to several prominent Black Lives Matter activists and organizers. I worked my way back through time and landed in the early 1990s, when the Web was born. I spoke to this person and that one. Many of them turned out to know, or know of, one another. Their voices and connections took me even further back in time, to the mid-1970s. In that journey through decades past, I encountered a group of people I call the Vanguard. They were black folks who, from the mid-1970s through the 1990s, used, built, and developed computing technology, digital networks, and online communities that furthered the interests of black people throughout the African diaspora.

Suddenly, my questions changed. I no longer contented myself to find the roots of today's digitally enabled, racial justice activism. The fact that I had myself encountered a whole group of people whom our Internet, computing, and media histories had never known—much less remembered—led me to ask a more fundamental question. What is, and has been, black people's relationship to the Internet and computing technology?

That's when I discovered a more sobering story. The story should not have been surprising, given America's racist history. Nonetheless, it was a chilling story to resurrect. That story was laid bare in civil rights, computer industry, higher education institution, and state and federal government archives, and periodicals from a bygone era. The story I found there I call "black software."

For me, "black software" conjures the myriad ways that we mobilize computing technology. Black software refers to the programs we desire and design computers to run. It refers to who designs the program, for what purposes, and what or who becomes its object or data. It refers to how, and how well, the computer performs the tasks for which it was programmed.

In this book I tell two black software stories. The first unfolds throughout book one. This is where the Vanguard begins to emerge. They are hobbyists. Entrepreneurs. Digital organizers, evangelists, activists, and knowledge brokers. The Vanguard positioned black folks, black content, and black culture to occupy the leading edge of the Internet's popular social development. Their collective stories begin in the mid-1970s and 1980s and take shape during the advent of personal computing, and the early days of personal computer networking. Their story extends through the World Wide Web's birth, and the dotcom boom's first bust. Their story demonstrates how black people have taken technology not originally designed with our concerns in mind. It is about black people using that technology to further our own personal, communal, and political interests.

But black software is also a story about how computing technology was built and developed to keep black America docile and in its place— disproportionately disadvantaged, locked up, and marked for death. This is the story that I unravel in book two. It speaks to the nefarious ways those in power use computing technology to destroy black agency by nullifying black people's hopes and dreams, aspirations, human potential, and political interests and limiting the heights we are meant to achieve. This story's shockwaves reverberate and still define our present day.

Book two begins in 1960, when students at MIT and other Boston-area colleges briefly aided their black counterparts in the South, who had launched the sit-in movement. This black software tale tells a story about American presidents, computing industry corporate titans, and the world's leading science and technology institutions. It is a story about the civil rights revolutionaries who forced white America to its knees, and the computer geeks who ushered in the computer revolution to lift white America back to its feet. It is a story about shadowy figures like those I call the Committeemen.

Black rage flooded America's streets throughout the 1960s. The Committee-men, at the behest of their government, set America's founding principles of white supremacy loose to run amuck in new computational systems they designed and built.

In between these two versions of black software—the kind that positively impacts black people's lives, and the kind that destroys them—lies a most significant question, not about recently popularized concepts like computer bias, or fair algorithms, or platform inequality, or digital ethics. No, the question goes to the heart of the matter that these concepts merely skirt around: will our current or future technological tools ever enable us to outrun white supremacy? After all, this is not just our country's founding principle. It is also the core programming that preceded and animated the birth, development, and first uses of our computational systems.

Book One

Chapter One

THE GREAT EQUALIZER

Black America dragged Jim Crow to its deathbed in the 1960s. Still, the nation's elite science and engineering institutions—the ones that were developing the first digital computers—locked black Americans out of their ranks. Educational institutions had not prepared most of these students to enter. And those who were prepared still couldn't get through the door.

That began to change by the time Derrick Brown came of age, late in the 1970s. His access to high-quality science and engineering education provided the first opportunity for him—and people like him—to change their relationship to the computer, and the men who once controlled its programming.

* * *

My first name, given name, is Derrick. Derrick means "leader of his people." Ever since I was young, it was taught to me by my parents what my name meant. If I tell you—no, I'm not an activist—I am denying my given name. My role as a leader is as a servant, through service. My service to the world is to be the great equalizer.

Derrick Brown arrived at Georgia Tech in the fall of 1992. At first sight, the campus community saw neither a leader, nor an equalizer. They did not see Derrick as he saw himself. Over the next two years, some of them would want Derrick to be something that he was not. Others would push him to do things inconsistent with his nature, with his name. But Derrick would always remain true to himself. Staying true was easy for Derrick to do. As far as he was concerned, he was the same person in 1992 that he was in 1979, when South Carolina's *Calhoun Times* readers first encountered his name.

* * *

As long as I can remember, I've been a writer. I love writing.

Derrick scribbled words on paper as early as he knew how. His third-grade teachers taught him how to write an essay. In fifth grade someone recognized that he was good at it.

My favorite activity is baseball. I've played baseball most of my life. It all started on August 23, 1974, my third day of kindergarten. I didn't know the rules of baseball or what a baseball was. I did know how to read well. The next day, we went to the school library. I checked out a book called How to Play Baseball. *I took it home, read it for a week, and came back to school all smiles. People always called me "shorty" and "pipsqueak," but I didn't mind.*

That day we played a game. I was the first man up. Usually the first man up is the one who can get on base quickly. I swung all the way around on the first pitch. I was a little disappointed. I waited for the next pitch and I hit a home run. I was proud of myself and gained more confidence in myself.

The next year I didn't play any baseball. I saved some Greenbax stamps and got a ball and a bat. A glove was given to me for my birthday. The next year I played as much baseball as I could.

Another year went by, and I ended up in a new school. I had forgotten how to hit. Although I did hit home runs and led in home runs by a sizable margin, baseball became boring.

Now in the fifth grade, I've changed my batting stance to four different positions. This has made baseball more interesting. I'm learning more about baseball every day. I watch baseball every Saturday to learn important plays. That is how baseball became my favorite activity.

Derrick continued to hit home runs on the baseball field as he did in writing. His fifth-grade essay won a South Carolina statewide writing contest. He repeated his accomplishment three years later, as an eighth grader, in 1982. That time the newspaper printed more than just his award-winning composition. His essay's subject was valorizing women in the workforce, and South

Figure 1.1. Derrick Brown (far right), receives award for excellence in expository writing. Courtesy of Derrick Brown.

Carolina's lieutenant governor personally conferred Derrick's distinction. In the *Calhoun Times* photograph that accompanied the essay, Derrick stood alongside another black boy, and a younger, but much taller, white girl who stood beside Nancy Stevenson. Stevenson would become South Carolina's first (and only) female statewide elected official.

As he entered high school, Derrick still churned out essays and other written pieces. He tired though, of his pen and paper, writing everything out by hand. He pleaded with his mom to find a solution.

Mom, we need a typewriter.

Well, we got a typewriter. Your aunt—she had a typewriter when she was in high school back in the 1970s. Use that.

Derrick instantly and eagerly complied. So quickly, in fact, that he'd forgotten something very important.

Mom, I can't type. I don't know how to type.

Well, you should take typing in high school.

Well, um, no. They have me on this college prep track. I'd love to take an extra math class. I don't want to take typing.

Take typing!
I don't wanna take typing.

Dogmatic and determined, Derrick won that battle. His mother, a kinder-garten teacher, was not going to argue with her son wanting to take an extra math course. She had another solution already in mind.

I'm about to get a computer. We're gonna get an Apple IIe. That's what I'm gonna get. And you'll be able to use it, because it has a keyboard on it, so you can use it just like a typewriter.
Okay.

Derrick and his mom traveled twenty miles west on State Highway 301 to Orangeburg to *play*—as Derrick described—with the computer. He thought to himself as his fingers stroked its keyboard, wondering how he could accomplish typing tasks on a computer without being able to type. *Even if I can't type, I can just look at the keys. The keyboard is nice and soft, not heavy like the typewriter.*

Mrs. Brown financed the computer. The Apple IIe cost upwards of $2,638 (that, plus the finance charge, would have run her close to $7,000 today). She had some explaining to do with Derrick's dad, but she had already done the deed. As he prepared to enter Calhoun County High School, Derrick wasn't just a two-time composition award winner. He was a member of the first family in town to own a computer.

It was my mom's computer. But I borrowed it. That's when I learned about software. I figure, the only way you interact with a computer is through program-ming, so I start messin' around with programming, just using the owner's manual for the computer. I found out that (A) I don't understand it, and (B) I don't really like it. And B is probably correlated with A. I don't like it because I don't under-stand it.

Frustrated, but not deterred, Derrick explained this to his mom. Again, she helped him find a solution to his problem. His mother enlightened him.

You can buy programs to run on this where people have written all these instructions that you're trying to learn how to program. Someone writes all those for you and you just put the disk in, and you use the program.

Derrick discovered AppleWorks. Apple had sold the software program at a discount to Apple IIe buyers. His mother didn't tell him how much the software cost. But Derrick was grateful to find out a solution existed. The software bundle included a word processing program. Derrick could have used it to further his writing, but he stumbled upon the database application instead. He had already become familiar with databases in school. He made up his mind then and there that it made more sense to buy software than create it himself.

Derrick scrutinized each program as though with a magnifying glass. He learned the secret powers each revealed. After he had created his knowledge base, Derrick reveled in getting the chance to show off his skills, focusing on something he loved, no less. In the summer of 1985, Derrick attended the South Carolina Governor's School for Academics. To prepare, each student chose an academic major for the summer school.

I'm going to choose computer science, Derrick thought to himself. *I'm gonna get to spend the whole summer building databases!* He was in for a rude awakening. The first day of class his professor announced that they would spend the summer writing code, in BASIC. BASIC is a computer programming language that stands for Beginner's All-Purpose Symbolic Instruction Code. Derrick's mind wandered back to that Apple IIe manual and his futile attempts to program his computer, his frustration, and eventual exasperation. Then, Derrick quietly raised his hand. He spoke softly. Confidently. But not enough to appear obstinate or demanding.

Professor, can I . . . uh . . . do databases instead?

Tell you what. You're going to do a project at the end. So for your project, yes, that sounds like a good idea.

That's great.

You can do whatever you want to do for your project. But in order to do whatever you want to do for your project, I want you to learn how to write this code.

Derrick appreciated the professor's flexibility. He did the work. And he was glad he did. In the end, he had learned something new. But that still did not change his mind much about programming. Derrick loved writing because it allowed him to bring together words on a page, in a rhythmic way. Those words conveyed his ideas. His words could offer inspiration and hope. *When I'm writing code*, he thought, *all I'm trying to do is stitch together the algorithm to score a bowling game. That's cool, but it's not the same as writing.* Nevertheless,

Derrick did the work. He learned what he needed. And he got to shine when that final project came around.

Databases were my thing. I liked to collect factual pieces of data. I used the database to make an all-star team based on metrics instead of popular booking. I took my baseball abstract by Bill James. I had all the numbers for everybody and I punched them all in, and made those filters to show that my all-star team had to have thirty home runs, a hundred RBIs and a .275 batting average. Had to have .350 on-base percentage, a .600 slugging percentage. That's how you get on the all-star team. Pitchers had to have a 2.50 ERA. You had to throw at least 200 innings, at least 175 strikeouts.

Database applications worked for Derrick because he really knew how to exploit them. He also knew baseball like he'd been playing, coaching, and calling games since he walked out of his mother's womb, wearing a baseball helmet and with a Louisville Slugger resting on one shoulder. Even without being an expert programmer at age sixteen, he realized that he could bend the database software application to his will. Why? Because he knew exactly what he needed from it, what he wanted to get out of it. He felt in control.

By the time Derrick had advanced through high school, he had developed a philosophy, parameters that defined his relationship to computers and to software.

I'm a writer. That's my lane. I want to take these tools, and make them beautiful, and make them do beautiful things. If I can find some beauty, some application, I'm going to grab that because that is what will compel other people to engage and use it.

The more Derrick played baseball, the more he wrote. The more he learned new things, built bigger databases, mastered the magic he saw in numbers, and recognized the value of data, the more Derrick got noticed. The next time someone sized up Derrick's diminutive frame and recognized massive potential, he was in the tenth grade. Someone was recruiting him—and it wasn't a baseball coach. A Clemson University professor of electrical engineering and computing wanted Derrick on his team.

At the time, the United States employed 1.3 million engineers. Only a few more than 35,000—2 percent—looked like Derrick. Around 1976, Dr. Robert W. Snelsire and his colleague Dr. Sue Lasser had started the Clemson Career Workshop. By 1986, when Snelsire first met Derrick, the program that he had spent a decade building was reaching its stride.

We bring the best and the brightest to campus during their sophomore year. Then we ask them back in their junior year. We give them intense preparation for an engineering curriculum. And then, of course, we tell them all the glories of Clemson. It works![1]

Sure enough, that strategy did work for Snelsire. And it worked for the university. By the mid-1980s, Clemson—a university nestled in the South Carolina foothills—wasn't just producing more engineering graduates than most other degree-granting institutions. It was producing a sizable proportion of the country's black engineering graduates.[2] The workshop, designed to attract students of color to engineering, also worked out quite well for Derrick. As Snelsire knew he would, Derrick made his way to Clemson for the workshop in 1985. He returned the following summer in 1986. And in 1987, Derrick knew that he could go to MIT or to Stanford. But he chose Clemson.

Clemson. They invited me. They said, "Hey, Derrick, you're from South Carolina. We would love to have our homegrown academic talent matriculate right here in our state. So come on up. And we'll pay for it!"

Those last five words were the magic words. In the fall of 1987, Derrick went back through the Blue Ridge Mountains to join Clemson Engineering Department's class of 1991.

To truly grasp the magnitude of Snelsire's, Derrick's, or Clemson's accomplishments, you must understand something. This was the engineering and computing field's second chance. That Derrick would master his computer's software was neither foreseen, nor planned for. Such revolutionary tools were not originally designed by, or for, folks like him. Black folks. Derrick's life would never let him and those like him forget that fact. And Snelsire's Clemson recruitment strategy wasn't a novelty. But it was very much a corrective to a problem first highlighted back in the early 1960s.

* * *

Back in 1964, Massachusetts Institute of Technology (MIT) faced a disturbing problem. At first, they did not see it, did not want to consider it, refused to own up to it. It had been there all along. But by 1964, the rest of the academic community had begun to notice a trend that not even MIT could ignore any longer. What was the problem? Was it their graduates? After all, society judges higher education institutions by their product: the students who graduate.

But MIT had no problem with its graduates. Only a handful of them left the institute at the end of the 1963–1964 academic year having not found a place to successfully launch their careers. And even that handful was probably doing just fine a few months after MIT had completed its end-of-year accounting. According to its placement bureau, nearly half the institute's 1964 doctoral degree recipients left the campus for jobs in private industry. More than a third of its master's degree, and a quarter of its professional degree graduates did the same. Meanwhile, three-quarters of its bachelor's degree graduates continued their studies. They would, presumably, follow the same professional pathway chosen by those who'd just achieved the advanced degrees they were poised to pursue.[3]

So what was it then? Did MIT's academic standing slip that year? After all, we also judge higher education institutions based on the quality of its faculty.

But MIT did not have to confront that problem in the slightest either. The MIT faculty, administration, and students accomplished more than their share of milestones that year. For example, the institute had launched a new *Center for Communication Sciences.* The center addressed issues ranging from statistics to information processing, linguistics, neuropsychology, and cognitive information processing.[4]

The institute's Joint Center for Urban Studies, which was financed by The Ford Foundation, celebrated its fifth-year anniversary that year.[5] During the 1963–1964 academic year, several of their faculty had published important books. Robert F. Wagner, a Harvard graduate student, reviewed each of them. Of Edward Banfield and James Q. Wilson's *City Politics*[6] he wrote:

City Politics *examines the structure of urban politics: the electoral system, the distribution of authority, the centralization of influence; and analyzes the forces and groups involved: reform, non-partisanship, businessmen, the Negroes.*[7]

Then he addressed Nathan Glazer and Daniel Patrick Moynihan's *Beyond the Melting Pot.*[8]

In Beyond the Melting Pot *Nathan Glazer and Daniel P. Moynihan challenge the very idea of the melting pot through an examination of the Negroes, Puerto Ricans, Jews, Italians and Irish of New York City. The book fails to convey the spirit and depth of commitment involved in the civil rights struggle. I wish too that they had placed more emphasis on the forces behind the rise of Negro extremism and the effect of permanent poverty, on the Negro's response to the compound problem of discrimination and unemployment.*[9]

These "experts" had begun to make "urban" a euphemism for the Negro, the Puerto Rican, racial and ethnic minorities, the poor—that is, societies' failures. Its problems.

If there was any downslide in MITs academic standing at all, it resulted from losing long-time professor Norbert Wiener that year. Wiener's theory, called cybernetics, jump-started and extended the new computer revolution's reach into virtually every aspect of society.

Despite his death, Wiener's work, ideas, and spirit began to live on through the birth of Project MAC. "MAC" stood for two things—Machine Aided Cognition and Multiple Access Computer.[10] The US Defense Advanced Research Projects Agency (DARPA) funded Project MAC, which already involved about a hundred MIT faculty and graduate students. Researchers in the lab that year studied a broad array of research topics, from automated engineering design systems, to natural language input for computers, to total modularity: a design basis for computer systems.[11]

While MAC researchers swam in these lofty research streams, the lab's first, most significant, and most surprising finding had been as profound as it was simple. *When more than one person can use a computer at the same time, those people will use the computer to communicate with each other.*[12]

Back in 1964, this moved scholars from talking merely about technological systems and relationships between people and machines to talking about information societies.[13] This was not an insignificant concept. Societies have citizens. Societies have politics. But at the same time the folks at the lab began to think about information societies, they failed to think about what kind of society they were building. Who was included and who was deliberately excluded from that society? They failed to think about, much less grasp, the consequence of building a new technological society in which Negroes were educationally and technologically both separate and unequal.

During that 1963–1964 academic year, MIT's faculty wrote and spoke a lot about "the Negro," and the complex urban areas where they lived. They built and studied new machines that facilitated new connections to people, and public and private institutions. But they never connected the two. No one asked how developing computer networking systems might impact so-called urban problems. Would it exacerbate them? Could the computer help solve them? One reason why they failed to do so has everything to do with the problem that was forced onto MITs agenda during that year. MIT's president at the time, Julius Stratton, explained: *We apparently have very few*

American Negroes in the undergraduate student body—possibly half of one percent of the total enrollment. Stratton received the criticism about as nonchalantly as a billionaire countenances a home electric bill.[14] He was not concerned, he was not moved, he would not change. Nor would James Killian, the chairman of the MIT Corporation.

It seems to me desperately important that at this juncture in the development of harmonizing sentiments in the US in regard to segregation in schools we exercise discipline and calmness and patience and that we hold emotionalism to a minimum.

Killian had made that statement just a few years before, when MIT hosted a conference about racial discrimination at American universities. Apparently the institute had decided that holding a conversation was enough to congratulate itself and move on.

MIT's predicament was a deliberate accident. Some of the circumstances that lead to it were out of the institute's control. However, the institutional policies that contributed to these circumstances were very much within its power. That year, MIT, like most US colleges and universities, was flooded with applications from women and men born just after World War II. It also received a record number of applications from what they referred to as "misguided students." The admissions staff deemed some of these applicants unfit for admission, or not genuinely serious about pursuing science or engineering. MIT rejected them outright. University officers also determined that some of these misguided students were simply poorly prepared. They had some natural ability, but they had not quite reached the academic threshold MIT expected of its incoming students. In prior years, the institution would have gambled on a few of those students.

But in 1964, they simply did not have the room. On the bright side, the university did double the number of women it admitted. Unfortunately, that made it even more impossible for MIT to admit any more Negroes than it already had.

As adamant as Killian and Stratton were about not tarnishing the institute's reputation, they could no longer fend off criticism from the higher education community. Stratton again explained:

The problem of non-academic factors in the admissions process is the most talked-about—and possibly the most critical—of all admissions problems in the country.[15] It ranges from the question of preference for athletes or alumni offspring to currently dramatized questions relative to minority groups. Traditionally,

MIT has insistently based its selection on objective evaluation of academic poten-
tial plus subjective evaluation of each applicant as a person without any specific
consideration for categories.

Geography and school were nonfactors, MIT officials claimed. Legacy
applicants received some "special handling." But none of them, they said,
benefited from special consideration. Meanwhile, *information about race,*
religion, and cultural background has been officially nonexistent and operation-
ally irrelevant. . . . In short, we have been unusual in the extent to which we have
refused to be influenced by categorical criteria or special-interest factors in select-
ing the freshman class.

MIT decided it could only do one thing to rectify its Negro problem. It
would attempt to send its admissions representatives to more "colored"
schools, and do so more frequently. The one thing MIT flat-out refused to
do was to reconsider any of its admissions criteria. The average student
admitted to MIT in 1964 scored in the ninety-ninth percentile on the
College Board's standardized math exam. The institute felt that any student
who could not hit that mark was not worthy to boast the institute's name.
From MIT's point of view, Negroes were, quite categorically, underprepared
to reach these higher echelons of the educated elite.

If MIT is to make significant contributions toward the education of minority
groups who now enjoy inadequate preparation in secondary school, it will have to
be done in directions other than the exercise of "reverse discrimination" in selec-
tion for admission and financial aid.

And so MIT decided that spring of 1964 that a single-digit number of
Negroes was as many as the institute could handle. The folks at MIT, and
those like them, were building a new information society. They, like officials
at most elite science and engineering institutions at the time, made the de
facto decision to exclude Negroes from designing, building, or deciding
what computing systems would be built. This would have devastating
consequences.

MIT changed its tune in 1968, when its student body and faculty unchar-
acteristically reacted to Martin Luther King Jr.'s assassination. That spring,
aerospace and aeronautics engineering professor Leon Trilling announced
that MIT planned to develop a number of pre-college preparation programs
to help recruit Negro students to MIT. Shirley A. Jackson—one of those
single-digit Negroes admitted to MIT in 1964—was in the room when
Trilling announced the new plan. Jackson was set to graduate that year. In

fact, she was set to become the first black woman to earn a doctorate from the institute five years later (in particle physics). Jackson told Trilling that *whatever MIT could do would be too little and too late.*[16]

Jackson was right. MIT's actions impacted a generation of Negroes who would never have a chance to take the reins of the new computer revolution.

* * *

Robert Snelsire must have looked back to the 1960s and recognized that a whole generation—his generation—of black Americans had not only been excluded from, but disadvantaged by the science, engineering, and computing revolution through the 1960s and 1970s. He was determined to do his damnedest not to doom another generation. Derrick was one of his beneficiaries. He was in a different position than any 1960s Negro as far as the computer was concerned. He was in the right place at the right time with the right skills and motivation to take advantage. What he didn't know was that he was preparing for the *next* computer revolution that was on the horizon.

* * *

Ellorree, South Carolina, was Derrick's home. Many of his town's nine hundred residents were family, of one kind or another. Those who were not blood were at least black. But for two weeks a summer, for two summers in a row, Derrick slept in a Clemson University dorm room, 155 miles away. There he learned to be a Tiger. He learned to be a good student. He started to learn to be an engineer. But in the fall of 1987, Clemson became more than just a summer getaway. It became home. And it felt nothing like Ellorree.

Clemson was built on land worked by slaves. After emancipation, black so-called convicts—practically slaves themselves—built the university's first buildings. They used stones from the old Calhoun Plantation.[17] Even in the 1980s, the spirit of Benjamin Tillman haunted the campus hall that bore his name. Tillman had been a notorious and vicious white supremacist, as well as a former South Carolina governor. He had presided over and ordered more than eighteen lynchings while governor and he claimed credit for several other race-motivated murders by other means. He once proclaimed that blacks *must remain subordinate or be exterminated.*

No one at Clemson in 1987 expressed the same sentiment. But a hint of it still permeated the school's culture. Midway through Derrick's first year,

Clemson's minority council said that its number one priority was *solving the problem of racial aggravation.* More specifically, a spokesman for the group at the time explained that,

The main group of incidents that confront us at the moment is the harassment of black resident assistants. This harassment takes many forms, including telephone calls at all hours, writing on the doors and walls, and vandalism of property. Another method is simply turning the music up real loud. A last straw was when somebody defecated in front of an RA's door.

These were apparently the kinds of things Derrick had to look forward to over his next four years: subtle and not-so-subtle signs that he was not welcome.

When Derrick picked up the campus newspaper, *The Tiger,* he saw many faces that looked like his. Most were athletes. Derrick idolized them. They were black like him. But, he still had trouble finding *his* people.

While he waited to find them, he had the good sense to go looking for Professor Snelsire. Derrick made his way to his office about as soon as he stepped foot on campus. He soon discovered that Snelsire was more than just a recruiter or someone who believed he had completed his job once Derrick enrolled in his first course. When Derrick ran up against a brick wall in one of his electrical engineering courses during his first year, Snelsire lent him books from his personal engineering library. And when Snelsire had Derrick in his own classes, he made sure to check in with him regularly. He asked, sincerely, whether Derrick and his classmates were on the same page. When Derrick encountered a professor who did not seem as invested in his success as Snelsire, Snelsire gave Derrick practical advice on how to manage and maneuver.

But even as Derrick welcomed Snelsire's support, the secret to Derrick's academic success had begun with the Clemson Career Workshop. It nourished Derrick's interest in science, technology, engineering, and math. Derrick remembered his days there very clearly.

We took a class in digital logic. I didn't understand anything that I was doing. But the guy that had taught us in the camp became one of my professors during my second year. When I took the class for real, he literally was teaching us the same stuff when we were in the eleventh grade that he taught us when we were sophomores in college. Second time around a lot of it clicked. I was like—okay, I've been so confused by this. But I get it now. And I know him a little bit. So now, if I don't know what he's saying, I know him. I can say—uh . . . uh, hey man, can you say that again? Knowing somebody enough to be able to tell 'em that you don't understand what they're saying. Now that's how you learn.

Before long, Derrick performed well enough to become a tutor in another program that Snelsire had started called PEER. He had designed the Programs for Educational Enrichment and Retention to help make sure that students like Derrick succeeded once they became Clemson students. Derrick performed so well that he became a respected and much requested source of academic support for other students of color in engineering and in PEER. That is how Derrick got to know other minority students on campus, aspiring engineers and otherwise. They flocked to him. They respected his academic skill. They required his wisdom about how to succeed at Clemson.

None of those students knew Snelsire like Derrick did. So they sought Derrick out, hoping to glean second-hand the mentorship and support that Derrick enjoyed directly from Snelsire. In fact, Derrick got so busy serving his fellow students that he started slipping in his own academic work. When he did, Snelsire, again, provided respite.

This dude gave me the keys to his office, man. Yeah, by the time I had become a leader in the recruitment and retention efforts on campus, he gave me a key to his office so that I could study, when I had to hide out. I was a tutor for the upper-level classes. So if I wanted to study myself, I had to hide. So he let me hide in his office, where he had a brand new Mac. That was a good thing. To be able to work privately in someone's office, where you have your own printer, instead of having to go into the computer lab, and dealing with the computer operators, who because they had access to the machines they thought they knew everything about everything, and if you asked them a question they'd look at you like you [are] stupid. So yeah, him giving me a key to his office, to go in there whenever I needed, that was above and beyond. And in a climate like the one we had at Clemson—it was hostile. It was hostile. But knowing that someone like Snelsire had your back. That meant a whole lot. A whole lot.

As their relationship deepened, Snelsire—who had become "Dr. Bob" to Derrick—became comfortable enough to ask something about the way black students interacted. For more than ten years Dr. Bob had prepared minority high school students to pursue engineering. For more than ten years he had ushered many of those same high school students to campus. For more than ten years he prepared them to succeed. Year after year he shook their hands as they walked proudly across a stage to clutch their well-earned diplomas. Yet even as he witnessed all of their successes, something still nagged at Dr. Bob.

Dr. Bob—at heart still an upstate New Yorker and Carnegie Mellon science geek—never understood what he thought was the peculiar way that black students like Derrick acted. After ten years, Dr. Bob had as close to a confidant as he'd ever had, in Derrick. Derrick understood Dr. Bob. *He was a little flippant. Very eccentric. But always sincere.* So when Dr. Bob asked, Derrick knew that he really wanted to know.

> Derrick, I have a question. For years I've seen . . . Well, something. I've never understood it. Look, can you tell me, um, why do all the black kids sit together in . . . in the lunchroom?
>
> *It's very simple. When I first came to Clemson, and I came into the lunchroom, all I saw was a lotta white students sittin' next to each other. And, they didn't look like they wanted to sit next to me. So I found the only other black person sittin' in the lunchroom, when I was there at that time of the day, and I went and sat by 'em and got to know 'em.*

That's it?

> *That's the answer to the question. That's it. I know you were expecting more, but that is the answer. I went and sat where I felt comfortable, and I kinda stayed away from where I didn't feel comfortable, and I kinda think that's what all of us do.*

By his junior year, Derrick felt at home at Clemson. He had grown into his academic strengths. He remained true to his name, helping fellow black engineering students overcome their academic obstacles. And, he had, finally, found his people.

It might have been a hostile environment, but it was the warmest environment that we knew, so we decided to take a chance, largely because we had already experienced it enough to know that there would be other people there to support each other and that we definitely would get a great education because we'd already been exposed to it.

But Derrick needed more. Many of the black students, including Derrick, felt that Clemson continued to foster an intimidating racial climate, one not conducive to learning or living. Derrick saw himself as a leader. He felt responsible to his fellow students, and not just for helping students like himself to feel comfortable. The times demanded change and he aimed to be its catalyst. As he prepared for his junior year in the fall of 1989, Derrick was determined to make good on the demands he placed on himself. He campaigned for a seat on Clemson's Minority Council.

Even more than fighting racism, discrimination and hostile treatment on campus, Derrick knew what other fights awaited him if he were to be successful. Earlier that year, Derrick had witnessed the Martin Luther King Jr. birthday commemoration event the Minority Council organized. Neither Clemson, nor the state of South Carolina at the time, recognized the day as an official holiday. Some black students skipped classes to protest. Others took part in the Minority Council's march. *Tiger* editors asked Markus Moore—the Minority Council's chairperson at the time—how he felt about the event's turnout. The editors also asked him why so few blacks in particular participated in the MLK and other campus events designed to challenge racism. His answer? Apathy.

The fall came. Derrick won his seat on the council. And as a leader on the Minority Council, Derrick was prepared to fight so that black students at Clemson could take full advantage of their education. But there was a problem. Derrick's fellow black students did not agree about just what fighting for one another meant. Derrick believed King's life and work meant that all black people had an obligation to uplift the others. One of his council colleagues thought it meant black people should take every advantage to advance their own interests.

By February 1990, everyone could see that Derrick was quickly learning to be an activist as much as he was an electrical engineer.

Whenever something would happen on campus, they would always run the same composite sketch in the school newspaper. And that person was always obviously a person of color, obviously male, and obviously the same person. I'm not joking! It was always the same drawing.

To draw attention to that, the black students got together and took the newspaper hostage. They learned all the drop-off points for the weekly distribution, and they just followed the delivery trucks. And every time the delivery truck dropped a stack, they'd pick that stack up. They got some megaphones and had a press conference standing on the stack of newspapers. All this to say, we at least need a better artist drawing these sketches. 'Cuz it's not the same person committing all of these crimes!

By the end of the month, Ms. Jennifer Brown became the first black editor of the *Tiger*.

Derrick, mind you, wasn't one of the students jumping in and out of the car, picking up newspapers. He didn't stand on a podium. He made no pronouncement to the crowd. That was not his way. Derrick may have wanted to engineer a revolution. But he never intended to be its frontman.

Chapter Two

THE TECH SCHOOL ROUTE

Before the 1980s, there were only two places to gain regular contact with a computer: the university or the workplace. Both closed their doors to blacks from the beginning. But by the 1970s, both institutions began to see how black people's presence benefited them. A steady stream of black folks began to flow through them. Some, like Derrick, engaged computers through university education. Others, like William Murrell, secured access through employment.

An engineer interacts with computer systems differently from the technician. But both paths gave Derrick and William opportunities to approach these systems in a way that worked for them—and us. Black people.

* * *

One way or another, William Murrell was going to do his own thing.

I was good at mechanical drafting. I was in high school, and I fell in love with electronics and wanted to become a recording engineer for studio and live performance. My mom said I had a knack for it because I was always fixing TVs, radios, and things. Every concert I went to I spent more time talking to the audio board mixers than I did watching the show. I volunteered at a local music store to learn stage set-up.

William wasn't like Derrick. College was not going to be his thing. Nor was it ever really an option. The university was fine—for other people. William was just not going to follow the same path. He wasn't familiar with that track. Nor was he particularly poised to pursue it. William had considered what he was good at. He looked around with laser-like precision and identified the kinds of people he wanted to emulate. That's when success for William—and the path leading to it—started to look different than it had for Derrick.

When a DeVry salesman made a pitch to my twelfth-grade drafting class, the audio engineering path sounded doable by learning "electronics." I had no "bachelor of science in electric engineering" people to compare messages with. Also the two-years-and-out tech school path and lower cost was very attractive. I knew no one who worked in engineering or electronics at the time except the roadies. None I talked to got into the business via college. They knew somebody who put them behind the audio board. They didn't go to tech school either. My dad was going to enroll me in RCA electronics tech school in New York where I could live with him.

William had already grown independent. He was born in Harlem, New York, in 1955. But William left Harlem and moved down south to Gastonia, North Carolina, when he was just six years old. His father stayed behind. The fifty-square-mile town that was home to about fifty thousand people gave William room to stretch, freedom to grow, and the opportunity to cultivate a mind and will of his own. Sure, moving to New York would have allowed William to pursue his dream in the comfort of his father's home. It would have provided him the financial cushion his dad was prepared to offer. But William chose to make it on his own.

William decided to go to Atlanta. And he worked for a year after graduating high school to make it happen. The textile industry was all that Gastonia had. So that's where William first found work. It was where he first remembered being rewarded for his ideas.

I worked at Firestone Textiles, which made thread for car tires. They had suggestion boxes for employees. I entered my suggestion to improve safety on a Criller Yarn Spinning Machine. It got picked. I was summoned to the office to receive a cash award and take a photo for the company newsletter. Check was $25, and they withheld taxes on that!

William had no shortage of ideas. And in the fall of 1976, they landed him at the DeVry Institute of Technology in Atlanta, Georgia.

Atlanta had this Allman Brothers band culture, with two cool black players on drums and bass.

Music attracted William to Atlanta as much as the satisfaction from knowing he was paving his own way. Atlanta was the big city, at least compared to

Gastonia. It displayed all the big city ugliness and problems: poverty, segregation, crime, discrimination. William confronted all of it at DeVry, though in relatively insignificant measure.

DeVry's main intake and orientation facility was a classy dormitory located on Peachtree Avenue, which ran through the heart of downtown Atlanta. DeVry's location appeared symbolic. Just one block away stood the South's most prestigious science and engineering higher education institution referred to as the MIT of the South: Georgia Institute of Technology.

DeVry launched its bait-and-switch from this strategic, Peachtree location.

In fact, the year after William started there in 1973, the United States Department of Justice slammed DeVry with a lawsuit alleging that it steered students to segregated housing.

The dorm they showed me was all black, no question about it. It wasn't a good studying environment. Facilities were poorly administered and maintained. I teamed with two guys and rented an apartment by the second week. Student life was integrated. But everybody had trouble with a group from Liberty City, Miami. I didn't see violence or drugs, just too much noise and trash talk. Dropout from start to finish was high. Not sure how high but the class I graduated with was much smaller than the number of freshmen who started.

Whether William realized it or not, there was a business model that produced the teaching and learning conditions he experienced at DeVry. In the early 1970s, the Radio Corporation of America (RCA), the Columbia Broadcasting System (CBS), Ling-Temco-Vought Aerospace Company, and other companies like them, pioneered and profited from a new business trend. They purchased and expanded profit-making technical training schools. For anyone willing to pay the price of admission, these schools offered the same technical skills training that they taught in-house to their own workers. The new schools invested in instruction, but skimped on everything else. That is what created the profit potential.

Bell & Howell, a long-time motion picture camera equipment developer and manufacturer, purchased DeVry in 1967. For a price, DeVry offered its students something that American colleges and universities were increasingly unable to guarantee. It offered the opportunity to make things, and so DeVry made it possible to realize the fruits of one's labor almost immediately. More important, to the successful student, it promised no shortage of positions in good-paying jobs.

Eight resident schools prepare students for the world of work. They utilize advanced teaching systems to train more than 10,000 full and part-time students for careers in electronics. Curricula for all resident schools' programs are evaluated and adjusted on a continuing basis to assure maximum correlation with the needs of industry. This includes identifying technological changes, product development and job demand in order to train students in appropriate skills.[1]

DeVry didn't invest much in its facilities, and its racial climate was fraught. But damned if DeVry didn't deliver exactly what it had promised William: an advantage.

Instructors were good, hands-on. The math teacher was a very short black woman, smart as a whip. But my math skill was lousy. And I wasn't confident about passing the engineer course once I became aware what an engineer does. But I was learning radio broadcast systems as part of electronics theory and heard about getting FCC class licenses to operate station equipment. I went to libraries in my spare time and studied this test. I took it, passed the first time, obtained an official FCC Radio Telephone Class B license before graduating. It was a specialized test—like passing the bar exam. That meant something.

It mostly meant that William was well prepared to get his first job. He wanted to stay in Atlanta. But in 1975, while William prepared for a new career, oil prices spiked, and the stock market crashed. The country's Bretton Woods monetary policy dissolved. Jobs still existed, but they just were not the ones William wanted. Not in Atlanta, at least. When he quickly tired of working at a grocery store, William replied to a newspaper classified ad hawking jobs with Duke Power Company.

Duke Power was North Carolina's private, statewide electrical utility. Just a few years earlier it had defended itself in a landmark race discrimination lawsuit that the US Supreme Court ultimately decided. The case established a new legal theory of discrimination called disparate impact theory. The theory held that policies and practices that appear neutral, but have a disproportionately negative impact on a protected group (because of race, sex, etc.), could be found to be discriminatory.

But by the time William applied, any such policy and practices that Duke Power might have retained had no effect on William's prospects. Duke Power's communications group hired him immediately. And he came with his own advantage—beyond the fact that he was black.

The FCC license was required to work on the base radios and microwave transmitters the company used to create its own independent phone and data network it used across the state. I worked for engineers who designed stuff in a lab and sent out schematics and plans for field installs in vehicles, nuclear plants, steam-based generation plants, etc. So often techs had to figure out and debug this "engineered" stuff before it would work. At Duke, a communications tech didn't wear [a] suit and tie. Engineers did. But our work was similar in many ways. There were engineers who did not possess the FCC license, so to my knowledge they could not service equipment that I could do legally.

When I got hired out of DeVry by Duke Power I felt triumph. All the things my parents told me about staying focused came true. Bought a cool foreign car. Immediately.

I knew friends, ex-high school football heroes, college grads who were not given a company car to drive and take home, and an unsupervised schedule to work four ten-hour days all across the state with hotel and meals paid for like my first job provided.

Even with all the excitement of working at Duke Power, by 1978, William was ready for a change. But he did not know that he was about to encounter a two-decades-long, twisted racial history at International Business Machines (IBM). That history first came to a head in 1968.

* * *

On Monday morning, January 15, 1968, the hearings commenced. The US Equal Employment Opportunity Commission (EEOC) had called New York State's one hundred top employers to account and atone for their guilt. The Civil Rights Act of 1964 had outlawed employment discrimination and created an independent federal agency, the EEOC, to investigate, discover, and eradicate all vestiges of such discrimination among the nation's largest companies.

The agency's first chairman, Clifford L. Alexander, opened the hearings promptly at 9:30 that morning. Anxiety filled Courtroom 110, at New York City's Foley Square District Court complex. Chairman Alexander wasted no time. He welcomed the audience and introduced his fellow commissioner, Mr. Vicente Ximenes, and his general counsel, Mr. Daniel Steiner. They proceeded directly to their business.

The Equal Employment Opportunity Commission has found from figures submitted to it by companies in the New York City area that two-thirds of the positions reported fall into the white collar category. We have also found that Puerto Ricans and Negroes have been excluded from employment in many firms and included only in insignificant number in many others.[2]

This was only part of the problem. The commission noticed something else in the data the companies had provided.

Several firms in New York were utilizing Negroes, Puerto Ricans and women at every rung of the job ladder.... Yet in an examination of the reports of the 100 major corporations producing approximately 16 percent of the Gross National Product, with home headquarters here in New York City, our report reveals the shocking and disillusioning fact that 48 of these 100 companies do not have a single Puerto Rican in approximately 8,000 official and managerial jobs.... It became regrettably clear that 56 of them did not have, among 12,000 officials, a single Negro Official or Manager.[3]

IBM was one of the fifty-six.

In October 1966 there were 61,170 Negroes enrolled in academic and vocational high schools in New York. There were almost 40,000 Puerto Ricans.... These young people must have the total opportunity they were taught to expect in a democratic society. The foundation of their opportunity is the ability to get a job according to their skills and, as important, to move all the way to the top according to merit, without being inhibited by the irrelevancies of national origin, color, sex or religion.[4]

The commission's report highlighted several other significant findings. The "total exclusion" of minorities from white-collar jobs was pervasive. Where they were included, blacks and Puerto Ricans fared better in technical and clerical roles, rather than managerial ones. Where women were well represented, black women generally were not. Minorities in white collar jobs were not paid as much as their white counterparts. And, lack of skill was insufficient to explain minority under-representation. The EEOC concluded from all this that *any explanation for these relationships must ascribe a role to discrimination.*[5]

Opening statements closed. The harsh realities of private-sector employment discrimination dangled in the chilly courtroom. For the remainder of the hearings the commission invited representatives from the one hundred corporations to explain their failings and describe their plans to do better. Metropolitan Life Insurance Company. Goldman Sachs. The *New York Times.* Columbia Broadcasting System. Shell Oil. Bristol-Myers. Representatives from the financial, media, communications, energy, pharmaceutical, and personnel industries all testified. No one from IBM showed up. As part of the commission's investigation, however, IBM had submitted a report for the record.

IBM prefaced its report with historical context. It pointed out that the company employed 7,600 people in white-collar positions in New York City. It explained that 590 of these were members of minority groups. It touted the number of women employed in high-level technical, scientific, and basic research positions (almost all of whom held at least a college degree). However, the report remained silent about how many blacks IBM employed. It did not take much, however, to read between the lines. Black employees were few and far between.

IBM's report explained that it started its equal opportunity efforts back in 1953. It reminded the commission that IBM had been one of the first of fifty companies to sign on to President John F. Kennedy's *Plans for Progress* initiative in 1961.

Nevertheless, IBM's report to the commission had made its point. The company had at least been trying, for some time, to increase the number of black and other minority employees among its ranks. And it marshaled more than just its nondiscrimination policy as evidence. It presented an articulated strategy to reach Negro employment goals, including the creation of a position called the director of equal opportunity programs. The first director operated from IBM's Armonk, New York, headquarters. Under the director's leadership the department produced what IBM called its "Blueprint for Action."

IBM referred to its outreach and recruitment efforts as liaison programs. IBM provided direct, unrestricted grants to historically black higher education institutions like Tuskegee Institute, Howard, Dillard, Fisk, and others. Sometimes it provided scholarships to students at similar institutions. And at others, it awarded fellowships or loaned equipment to members of the faculty.

Other such efforts aimed to impart new skills to existing workers. IBM executives instructed its managers to motivate minorities to take part in the initiatives. But it had been standard policy that applicants had to take an aptitude test to qualify for such additional training. In fact, IBM had developed the first such test in the industry, what it called its "Programmer Aptitude Test." More than anything else, the test itself, and its widespread use, reinforced the idea that computer programmers were born not made. Blacks who failed to pass the test were seen as unteachable.

An IBM task force is presently assessing the company's overall testing program to determine whether, and to what extent, aptitude and achievement tests may be restricting job opportunities for minority-group applicants. In the interim, IBM management at all levels have been specifically advised not to use tests as the significant measure of an applicants' qualifications or potential.[6]

While it put testing on hold, IBM's "Blueprint" efforts began to publish job openings far and wide in black communities scattered across the country. It employed liaisons from black community organizations like the National Urban League, who could exploit their connections to black newspapers and magazine outlets for IBM outreach purposes. Papers like *Amsterdam News* and magazines like *Ebony* distributed IBM content in New York City, for example. *The Defender* served IBM's purposes in Chicago, as did the *Baltimore Afro-American* in Maryland. IBM didn't want this to seem merely like company-sponsored advertising. It worked hand-in-hand with trusted community leaders who connected their constituents to all that IBM had to offer. At least that was the appearance IBM strove to cultivate.

"IBM Expanding Research, Demands Huge Workforce" read one exemplary headline in the *Baltimore Afro-American*. The "story" introduced and summarized IBM's desire to expand engineering research and development. The several-column-length story listed the job categories IBM looked to fill. It described each one in great detail. The list began with the lowest category and continued up the hierarchy. "Technicians" occupied the lowest category on the corporate ladder. They called these people "customer engineers."

Customer engineers worked directly with customers who had purchased and worked with IBM equipment. They initially installed customers' hardware and software products, and they trained the customer on how to use them. They also provided ongoing troubleshooting when necessary. Some technicians' jobs were less customer centered, like mechanical, electromechanical, physics, or chemical technicians. People in these positions provided technical assistance to research scientists and engineers.

Draftsmen and designers occupied mid-range position categories at IBM. The former was a gateway to the latter, and required specific technical knowledge. Scientists and engineers sat atop the IBM career ladder. These included researchers, development engineers, product engineers, product test engineers, and programmers. As the liaison newspaper and magazines made clear, these positions required advanced math, science, and engineering education and training.

Young men who have the aptitude and interest . . . will have a rare opportunity to take a real "under the covers" look at the IBM company this Saturday since the

Chicago Urban League is now conducting a campaign to help IBM find more talented college graduates for jobs in all its major areas, today this column will deal with the work of the programmer.

At the time, Hampton McKinney was the director of the Chicago Urban League's Employment and Guidance Department. He had drafted this call to be featured in the *Chicago Defender*. Clearly McKinney sourced his words directly from IBM. Nevertheless, McKinney crafted the invitation.

Talented men and women college graduates to consider applying for a job working with one of the most complex tools ever created by man. Programmers start with a statement of the problem. They study and analyze it. Then they organize their information into a logical sequence of facts that can be developed into a step-by-step procedure to reach a solution. The procedure is then coded into a language and form which can be handled by a computer. IBM programmers delve into every aspect of the programming art: systems programming (including control programming, language processors and other techniques), scientific problem-solving, commercial and control applications programming and others.[7]

This was the earliest and simplest way for IBM to say that programmers create software. This was a vaunted role at IBM. Software engineers were decision makers. They directed the computer to make certain decisions in line with customer needs. More important for IBM, software developers helped decide what their customers needed. And that need was limited only by what the programmer could build.

IBM's outreach to black and minority communities appeared to respond to the federal government's new equal opportunity demands. However, IBM's heavy-handed description of the jobs it offered made it clear that it had merely substituted its aptitude tests with job descriptions it knew would weed out a large proportion of potential black applicants.

While IBM did try to find what unicorns it could, its primary black outreach philosophy focused on what IBM referred to as "supply-side" recruitment. IBM developed its own talent, preparing people in its own way to carry out the company's own prerogatives. IBM provided technical training to high school students. Then it tracked and pipelined them from school to IBM through its "pre-employment or pre-assignment" training programs. The New York City Department of Education, for example, selected fifteen black students for a three-week, IBM Data Processing Division training program. Three years straight, its "graduating" students went directly to work for IBM customers. On a different occasion, IBM's Field Engineering

Division linked up with the National Urban League to develop an on-the-job-training program that aimed to *develop people who lack sufficient technical education for a customer engineering career, but who possess all the necessary aptitudes.* The program, IBM reported, had graduated 528 women and men since 1965. Of these, 211 of were black.

IBM extended its reach into black communities. But it, and companies like it in the expanding computing industry, publicized these positions at a time when the college and university science and engineering pipeline produced *very few Negro* graduates. Yes, IBM needed scientists, designers, engineers, mathematicians, and physicists. But the reality was that the nation's top scientific and engineering minds, pipelined through its elite science and engineering universities, had already flooded IBM's scientific ranks.

IBM was expanding exponentially toward the end of the 1960s. In just five years from 1965 through 1970, the company catapulted four positions to occupy number five on Fortune 500's company list. In those five years it doubled its annual revenues from 3.5 to more than 7 billion dollars. Orders for its Systems/360 and 370 machines created the greatest equipment backlog the company had ever seen. Its ranks swelled by 100,000 employees.

IBM was a decentralized company. But it had a pyramidal structure. Its relatively few executives presided at the top, and its prized basic science and research and development teams—its scientists, designers, and engineers—filled out its middle. But technicians were the company's support base. That's where IBM needed bodies. That's where it needed a massive population of technical laborers. That is where black people in the 1960s and 1970s were most prepared and directed to enter.

By 1968, IBM's president, Tom Watson Jr., had agreed with Bobby Kennedy that the company could meet its demand for technical and manufacturing labor and fight black poverty simultaneously. Watson devised a plan to invest in one of New York City's worst urban ghettos, Brooklyn's Bedford-Stuyvesant neighborhood. Watson thought Bobby Kennedy was an awkward conversationalist, with an odd interpersonal style. Nevertheless, on one of their few personal meetings, Kennedy offered to do Watson a favor. It was a show of kindness for the same kindness that Watson had extended to his brother John, and the Kennedy family. Watson needed new legal counsel for IBM. He especially needed someone who could help IBM defend against the steady stream of antitrust lawsuits that flowed its way.

Kennedy recommended either of two men. The first was Nicholas Katzenbach. Katzenbach had taken over as the attorney general after Kennedy stepped down to run for the Senate and then pursue the presidency. The second was Burke Marshall. Kennedy had—at first reluctantly—brought on Marshall to head the Civil Rights Division of his Justice Department.

Ultimately, both men joined Watson at IBM, and both men were critical to moving forward the Bedford-Stuyvesant project once Bobby Kennedy was assassinated in 1968. Watson explained:

The only time I ever saw [Bobby Kennedy] attract business backing was in 1966, when he decided to tackle Brooklyn's Bedford-Stuyvesant ghetto. He persuaded both Mayor Lindsay and Senator Javits to pitch in, and then recruited a bipartisan team of businessmen including me . . . and others. The white and black groups put together a combination of new jobs, housing renovations, and social services that gave Bedford-Stuyvesant a bit of new life. During the "long hot summer" of 1967, when there were race riots in dozens of cities, Bedford-Stuyvesant was quiet, and Bobby deserves some credit for that.

IBM made the biggest contribution of all to the effort in Brooklyn: we put a new plant there. Many of my business peers still thought that in America if you worked hard you reached the top. But it was obvious to me that in the ghetto, if you worked hard, the chances were you'd still be at the bottom when you died. We leased the biggest building we could find . . . on Nostrand Avenue in the very center of Bedford-Stuyvesant. Our plan was to employ five hundred workers—not as unskilled labor, but in real IBM jobs, with good wages, benefits, training programs, and the chance to advance and even get out of the ghetto by transferring to other IBM plants.[8]

By 1969, the plant was fully operational, manufacturing external computer cables and power supplies for IBM machines. By 1972, the plant produced a significant share of IBM's cables and power supply stock. It received a *Developing Human Resources Award* from *Business Week Magazine* for the more than four hundred Bedford-Stuyvesant residents that the plant employed.[9] By 1979, IBM Brooklyn manufactured a whole host of IBM products, including the IBM 3270 Information Display System and IBM 3814 Switching Management System.

The Bedford-Stuyvesant plant became a success. It demonstrated that investing in minority employees and communities could create real business advantages for IBM and its shareholders. By 1978, IBM was bursting at the

Mattie Arnold, Dept. 292, solders wires to a heat sink, one of the major components of a mid-pac. The manufacture of mid-pacs, a voltage and current regulation unit, began in January of this year.

Figure 2.1. Courtesy of International Business Machines Corporation, © 1969 International Business Machines Corporation

seams. It needed, again, to expand. That year the company broke ground on a new manufacturing and product research and development lab facility down south, in Charlotte, North Carolina. There, William Murrell was about to join the army of black technicians that IBM had spent years trying to produce.

IBM was successful—at producing technicians. It became even moderately successful building a black management class of marketing managers, plant managers, and public relations managers. Still, very few black IBM employees—technicians or managers—were placed in routine, much less key, science and engineering, programming, or research and development positions.

But William was ready to do his thing, in the position IBM believed he was best positioned for.

* * *

I wanted to wear a suit.

About the time that William felt he'd conquered his first job and needed a new, more rigorous challenge, IBM had just set its sights on Charlotte. The company had scouted the location for some time. On June 29, 1978, the *Charlotte Observer* announced the new plant. *It Adds Up—IBM Picks Charlotte*, the headline read. IBM had bought up space for a 600,000-square-foot complex toward the city's northeast side. It was strategically located near the University of North Carolina at Charlotte, in an area ripe for commercial and residential development. And this was key: the majority of the plant's first 1,600 employees were to be relocated from IBM's headquarters in Endicott, New York. IBM began hiring a relative handful of local employees that same summer and late into the fall. William was one of them, hired in IBM's already-existing System Communications Division.

I contacted a recruiter. He got me an interview with IBM in 1978. I was hired after that interview. I know the headhunter never got paid from IBM. He said they told him they had a "policy" of no recruiters.

IBM didn't need recruiters. Intermediaries, maybe, but not recruiters. It was IBM. And besides, it had been perfecting its minority recruitment game for more than a decade. Since the late 1960s, it had built multiple pipelines from black communities into the company's open doors. By 1974, IBM had made significant progress, but its president, Frank Cary, told IBM's board and stockholders that it was just getting started.

I'm sure you're all aware of IBM's commitment to Affirmative Action in providing equal opportunity for minorities and women. We've made good progress on one of our objectives—bringing into IBM capable and highly motivated minorities and women. Our second objective is taking longer to achieve: helping minorities and women qualify themselves for advancement at every level of the business consistent with their abilities and their growing population in the company. The relevant question I'm asked most frequently by IBM managers is: "How can we do that without practicing reverse discrimination?"

My answer is that we will not compromise our policy of promoting the most competent, most qualified people. But what we all have to do as managers is provide whatever extra help and learning opportunities may be needed to shorten the time necessary for minorities and women to compete on an equal footing with other IBMers. The best individuals will still be selected for promotion, but we intend to make the competition keener. We have tripled the number of minorities and women in the ranks of management over the past five years. That's not a bad start, but I'm convinced we can do better.[10]

For the first time, in 1979, IBM began to loudly tout not just its plans for progress, but its affirmative action credentials.

Emphasis continued in 1979 on worldwide IBM programs that provide equal opportunities in employment, advancement and training. Of the more than 16,000 IBM employees hired in the United States during the year, some 36 percent were women and 19 percent were minorities. At the end of 1979, the company had 2,000 women managers and 2,200 minority managers in the United States. IBM employees are presently on loan as teachers at 28 predominantly minority colleges. Since 1971, the company has loaned more than 230 employees for this program.

By 1979, William had become one of them—an IBMer, a customer engineer. The company, he soon found out, had quite a different culture from Duke Power. And, in many respects, his DeVry electronics training often put him at a disadvantage.

IBM was different. They strongly believed that if you were "teachable with the right stuff" they could make anybody a customer engineer. Case in point: My orientation was one-week learning IBM history and culture. Then they shipped me to a 90-day Washington, DC, system tech training school. They owned residences in DuPont Circle where we lived. Class was held downtown from there. My class cohort was thirty-five people. Not one person in that class had prior tech troubleshooting skills. Corzell, my roommate from Atlanta, was a saxophone musician. My lab partner, Fran, was a religious nut from Framingham. She was deeply religious—spent all her spare time in the Bible. Corzell got a girlfriend from the class from Detroit. She was a shoe salesperson. The training was sales/ customer service skills stuff. We acted out roles of customer personas.

Fran outscored everybody.

The tech stuff was hard. Real machines that had come out of the field with problems for us to find. I did bad on the test when using my experience and intuition because the troubleshooting exercise training was about following written decision trees and branches. My nut lab partner outscored everybody.

When you went on-site but the problem remained unresolved after a set time, say one hour of troubleshooting, the tech had to escalate to a dispatch center, then the cavalry (two senior techs) would come to bail you out and fix it while you watched and learned. The minute you escalated from the field you were ordered to stop work and wait. We were required to remove all customer software (it was on removable cartridges) and work entirely with our tech diagnostics we carried for every model. When the cavalry arrived they checked how you used the decision tree to begin the fix. That tree was surprisingly effective. When I would see an obvious way to fix something—it didn't matter unless I got to that point from reading the troubleshooting decision trees.

Every computer at the client's office had a written maintenance log. A record of prior history was important. It helped determine what troubleshooting branch you started at to fix the thing. Fixing computers was board swap but you needed the right electronic board. Fixing printers was heavy in mechanical gears, electro sensor timings, that sort of thing. These were minicomputer 3270 series that connected over three hundred baud modems to model 370 mainframes.

Unlike his counterparts from the decade prior, when IBM trained William, the customer engineer was fully networked. William received new alerts for service and communicated with other members of his team, as well as clients, using a computer-aided dispatch system. The system had been first designed and tested for use in law enforcement operations in the mid-1960s. But by 1978, IBM had begun marketing the system it had been using and perfecting for a decade. IBM trademarked it as the Personal Office System.

It's a natural progression. First came the postal service, then the telegraph, the telephone and telecopying. VNET [a network launched in 1972 and used by more than two hundred thousand IBM employees; PROFS was an application that ran on VNET] *allows IBM employees around the world to talk to each other almost instantaneously. It's part of the company now.*[11]

By the early 1980s the system became a feature across the organizational spectrum, from the White House to most large corporations like IBM. The system included some very key components and capabilities, including real-time messaging; information distribution; information search, retrieval, and storage; and calendaring and time monitoring. All of these features had a history—all not entirely favorable to black people. The computing features sprang from experiments and innovations made throughout the 1960s when the nation grappled with massive civil rights protests. But each of them also had a future. The difference between their past and the present, in which William operated, was that he—and those like him—knew the system. They even had some control over it.

For William, there was a downside to IBM. The company said that it despised conformity and groupthink. But William discovered that IBM had its own way of doing things. And it expected its customer engineers to do it IBM's way, not whatever William's way might be at any particular time.

Conformity was very important at IBM. You couldn't wear a sports coat with pants. Had to be a matched suit. Black, wing-tipped shoes were standard. Believing anything the boss told you was expected. IBMers were like white-collar professionals. Our tool boxes looked just like executive brief cases. Every man, no exception, shaved their face, except me. Only guy in the building with a beard. All the IBM bosses were middle-aged, fifty-something white men.

We spent a lot of time in the IBM office so professional etiquette went with the dress code. A lot of fake conversations happened in this office environment. The field was cool. Customers were fine and hospitable but I think IBM staffers were stuck up. IBM coworkers in the field services division had challenges creating sustainable friendships. I think that was because everybody wanted to be promoted. Competition among us was a thing. Perhaps the escalation system often pitted CE against CE while it was designed to be mentor-like. It wasn't. Human nature caused some techs to feel superior to others.

I quit voluntarily over a bad call my boss made in a "client down" situation where I was expected to back up a series of lies he told them over a mechanical replacement call. When I quit he was shocked. I told him it was because of the repeat lying. He said, "Don't repeat that outside this office" and "You are forever barred from working in an official capacity with IBM Corp."

In 1983, William's run with IBM came to an end. As far as he was concerned, he had triumphed yet again. He had worked for the computing industry's leader. He came to know every inch of IBM's hardware. He intimately knew its software. Being able to troubleshoot meant he could also reverse engineer computer processes and programs, and make designs of his own. He also had service skills, marketing skills, and business skills. He got it all from IBM, and now they were his. He was about to put them to work for himself.

Chapter Three

THE ROXBURY SHAKE

The handshake. That's what computer networkers call that moment when one computer connects with another.

A new kind of black software began to take shape in the early 1980s. It started with a handshake. A connection. Not between two computers, but between a computer and the black community. Derrick had made that connection when he realized that he could and should use his engineering knowledge and computer application prowess to help other black brothers and sisters achieve their dreams. William would have to leave IBM before he could make that connection. He would have to travel to, and settle in, a place steeped in bitter racial politics and racial justice activism. William made his handshake. He found a way to use the skills that he developed at IBM to connect and uplift his community.

But black folks like William had been late to the computer game in the first place because of another handshake that never happened. The architects of the new computer revolution once had an opportunity to connect their technological powers to the racial revolution stirring among black people in the South. They failed. And it had devastating consequences.

* * *

William was done with IBM. Frankly, he was done with Charlotte. He didn't leave immediately. He took a little time off to do other things he loved.

He helped friends open a vegetarian restaurant. He threw himself into his music.

I had bought a guitar while at DeVry. I struggled to learn how to play. I had been using Berklee College of Music self-teaching lessons all the while at Duke Power and IBM. I chose to focus on Coltrane-like jazz. While in DC I hung out in jazz pubs, learning. A musician convinced me to go where I could hear "the language" often.

Charlotte didn't speak that language. So William made up his mind that he was ready to follow the music—and a new job—anywhere but.

I was interviewing in Virginia, Iowa, and probably others when Ohio Nuclear, Inc., said, "Work for us and we will send you to Boston." I'm like, "Great. That's where Berklee is." So I'm in Ohio training in Computer Axial Tomography (CAT brain and whole body radiology scanner system controlled by DEC PDP 1134 computers) systems as [a] field engineer—these things cost $1 mil each—when the call came to move my sister.

I rented a U-Haul, drove up to New York, and moved my sister's belongings to her MBA school. Harvard. I had two days before I had to get back, get back to Solon, Ohio. I used my free days after moving her in to scan Boston streets. Boston clicked after I saw it myself. And Ohio Nuclear's offer to "move everything" for free and provide a new car upon entering Boston was a good deal.

William drove his car into the heart of black Boston. Atherton Street straddled Boston's Jamaica Plains and Roxbury neighborhoods. The poverty in Boston concentrated in four, mostly adjoining black neighborhoods: Jamaica Plains, Roxbury, Dorchester, Mattapan. The area's structural inequalities produced drug addicts and thieves, hookers and hoodlums, welfare cheats and skid-row bums.

Densely packed as it was with the so-called "underclass," black Boston didn't fill its streets with people who lacked the will to work or the wherewithal to succeed. There were places where almost everyone was employed, as well as blocks and corridors where the high school graduation rate and number of college-educated residents outpaced Boston's average.

But to William, life in Boston appeared similar to life under apartheid. Black Boston's educated middle-class folk often shared the same fates as Roxbury's less fortunate. White people lived in neighborhoods where fewer people worked, finished high school, and graduated college. Yet those people could still count on making more money than educated black people. They were guaranteed to live among fewer people below the poverty line, and enjoyed greater career mobility than black Boston's better-prepared citizens.

But William's chance relocation would prove to be life defining. He had one foot deeply rooted in preserving and promoting black identity and community uplift and another foot afloat in the computer revolution's rising tide.

William lived just three blocks from State Representative Mel King's campaign headquarters, in Boston's South End. Black Boston had long established itself as fiercely activist, and King exemplified its long history of civil rights struggle. King had led some of the area's most strident civil disobedience activities in the mid to late 1960s. Since then, however, he had become more of a political pragmatist than Black Power radical. In the mid-1970s his deep ties to the community and organizing skills helped him win a seat in the Massachusetts Legislature. But King had something bigger to accomplish for his black Boston constituents.

King reached out across the Charles River from his black Boston ambassador's seat. He migrated toward Cambridge's elite, white, and comparatively wealthy science and engineering community. That's when he extended his hand to MIT, ground zero of the new computer revolution, but an institute that still admitted very few Negroes. Yet it was also an institute that decided it wanted to reverse that pattern, and had designed and built the kinds of pre-college training programs that Derrick would later take advantage of at Clemson.

The collaboration King established was MIT's second opportunity to foster a meaningful relationship not only with black Boston, but black America.

They had squandered their first.

* * *

Raleigh is a handsome town; its main street, dominated by the state capitol, is wide and spacious; the storefronts are plain, not gaudy. In front of four of those stores some twenty Negro students were picketing. As a Northerner I expected, and felt, the tenseness of the city. The day before there had been a fight on the picket lines and a Negro boy had been hit with a tire chain. What I found there, by talking to Negro students and visiting their colleges, was a spirit and a method of action which made such incidents…incidental. Dangerous they were—and are—but they are not the key to the sit-downs. For the Negro students, like the earlier Montgomery bus boycotters, are engaged in a new kind of political activity, at once unconventional and nonviolent.[1]

On February 1, 1960—one month before Michael Walzer penned these words for the pages of *Dissent* magazine—North Carolina A&T students had walked into Woolworth's, sat down, and eventually shut down the store's lunch counter in Raleigh, North Carolina. As a matter of policy, the establishment refused to serve Negroes. Woolworth's was still closed when Walzer, a Brandeis alumnus and Harvard graduate student, traveled down from Boston to witness the protest. He returned home with stories of bravery, heroism, and strategic ingenuity. Those stories awakened Boston's student elite.

On February 27, 1960, more than 150 Boston-area students picketed six Woolworth's locations throughout the city, from Brandeis to Boston University, Harvard to MIT.[2] The students poured in, more than anyone would have imagined. After all, those northern kids had the privilege to ignore, let alone act to support, the southern Negro's struggle.

Boston's local chapter of the Congress on Racial Equality (CORE) hardly boasted fifteen members. But the southern demonstrations revived them.[3] Operating from Roxbury's Blue Hill Avenue—a little more than a mile away from where William would relocate—CORE organized the pickets. CORE designed the slogans students hoisted into the chilly, Boston air.

We Urge Woolworth to Change.
End Eating Bias in South.
Support Southern-Student Protest.[4]

CORE's leaders instructed the students to remain silent, even when tormented by onlookers. Bystanders called them everything from coon lovers to communists. CORE trained the students not to run afoul of the passive, nonviolent resistance principles that governed CORE and the broader sit-in movement. These white students were allies; their role was to follow orders, not make them.

But some MIT students pushed back against the group's tactics. It was not that they did not know their role. Neither did they object to it per se.

The same evening that CORE staged its protest, MIT's student-run campus newspaper—*The Tech*—published news about it. MIT students called for a meeting on campus. The question on the table wasn't whether to continue, but how to most effectively organize their future actions. By the end of that week in March 1960, MIT students joined their partners at Brandeis, Boston

University, and Harvard. Under CORE's leadership, they birthed the Emergency Public Integration Committee—EPIC.

* * *

From its start, MIT struggled to distinguish itself from Harvard (Brandeis and BU were never competitors). MIT's founder, William Rogers, and his associates touted the need to develop a new form of education. Their goal was to connect basic science to engineering practice and direct it toward real-world application.[5] Harvard, and most colleges of the day, built their curricula around the liberal arts. But MIT wanted to be an institution at which its motto—*Mens et Manus* (mind and hand)—was also its mission.

MIT existed to *produce both cutting edge scientists and engineers* who would apply scientific knowledge.[6] Like Harvard, MIT was founded, organized, and operated as a *corporation*. It established its connection to industry from the beginning. But MIT was also one of the first land grant institutions. Thus, at the close of the Civil War, it embraced the responsibility to uphold the nation's public institutions and promote the public interest.

Chairing the MIT Corporation in 1960, Dr. James R. Killian Jr. pledged fidelity to the institute's mission.

The industrial revolution has been superseded by a scientific revolution... and industry and government must develop a suitable environment to provide optimum opportunity for creative people to produce change and progress such that it properly recognizes scientists and engineers in their role as scientists and engineers.[7]

MIT's president, J. A. Stratton, followed suit.

We are concerned with the advancement of pure science.... Through the profession of engineering, we prepare our graduates to apply the results of scientific discovery to immediately useful purposes.[8]

Like Killian, Stratton recognized the dawning technological revolution. He saw a place for MIT at its center.

We believe it now to be our principal responsibility to prepare the student to cope with—and ultimately, indeed, to lead—a technological revolution that is proceeding with gathering momentum.... We must exploit the capability of modern computers.[9]

Killian believed that new theories of communication and control being developed at MIT would significantly advance its mission to be the new technological revolution's vanguard.[10] As the 1960s began, MIT's administration and faculty remained steadfast. It policed its territory and stake in the dawning technological revolution. It built a metaphorical wall to separate its intellectual pursuits from a civil rights revolution that was beginning to storm it. MIT's students, those working with EPIC, aimed to break down that wall. They faced immediate opposition.

For some time now, we have been following with interest the activities of a group which calls itself the MIT EPIC. We do not question the avowed purpose of the organization. If individuals believe in a cause and decide that the way to effect their ends is to picket Woolworth's in Central Square, this is certainly up to them. We regret, however, use of the MIT EPIC.[11]

From the beginning, students challenged MIT EPIC's very existence, initially in an unsigned editorial from *The Tech*'s editorial board. Opponents thought EPIC was subversive. But they did not challenge its motivations. Nor did they challenge the underlying cause for which EPIC fought. The collective opposition cleverly shifted grounds and objected on procedure.

EPIC is assuming the title and privileges of an MIT activity prematurely. We suggest that the leadership of this group acquaint itself with the mechanism for obtaining Institute approval and seek such before it represents itself as "MIT" EPIC.[12]

Other students on the newspaper's editorial board cast suspicion on MIT-EPIC's alliances. To get to the bottom of the organization's mission and motives, Barry Roach—MIT class of 1961—interviewed Harvey Pressman. Pressman was a Harvard graduate student and one of Boston EPIC's leaders. Apparently, Roach didn't buy anything that Pressman tried to sell him.

As a southerner and moderate integrationist, I cannot help but be disturbed with the recent activities of MIT-EPIC. I feel strongly that the MIT student is being misled as to the scope and purposes of the area-wide group of which MIT-EPIC is a fellow traveler.

Roach castigated the group's constitution, pointing out that at other EPIC-affiliated campuses, the group operated ad hoc. They were not a formal part of the institution. EPIC boycotted, petitioned, and raised funds for students expelled from southern universities because they resisted racial discrimination. They documented such discrimination, particularly in housing and employment. They supported a strong federal civil rights bill. But Roach cast aspersions on all of those activities by directing suspicion toward EPIC's organizers, which included CORE, the National Association for the Advancement of Colored People (NAACP), and labor unions. Roach was skeptical of both EPIC's tactics and its alliances.

In light of the above facts, I urge that MIT-EPIC seriously consider what they are getting into and either petition for recognition as a permanent activity or else disaffiliate entirely with the area EPIC organization.[13]

The students representing MIT-EPIC did nothing of the sort. The following weekend, they, and their compatriots, tripled their prior picket. One hundred fifty of them had shown up the week before; this time, 450 of them came out in force. Instead of picketing six Woolworth's stores, they targeted eleven.[14] By this point, MIT-EPIC had created a fault line, widening by the day, within its student body. Richard Neel Sutton, MIT class of 1962, lashed out.

As a member of the MIT community, I object to the use of my name as part of the title of an organization as outwardly prejudiced as any KKK'er ever was. We have been attacked because we are Anti-Riot. The difference between "student demonstration," as used by Southern negroes, and riot, that which follows "student demonstrations," is largely a matter of name only. Is it fair to picket Tech Drug because the coop won't sell somebody a coke? Is it justice if a group of individuals attempts to embarrass the local Esso dealer because his counterpart in East Podunk Tennessee won't sell gas to somebody?[15]

Jean Pierre Frankenhuis, MIT class of 1961, saw it differently.

The issue of integration has awaken[ed] the so-called apathy of MIT students. MIT-EPIC proved that this generation is in trouble, and doesn't seem to know it; for this I think we should thank it.[16]

Frankenhuis was a native of France. He had grown up in Brazil and majored in math. He didn't think MIT-EPIC was particularly ideal. And he was convinced that many of his student colleagues joined the organization just because it was chic to fight for a cause. But he gave them their due.

In the end, MIT-EPIC went all in. Michael Levin, class of 1962, became the group's secretary. On March 25, he penned a letter to *The Tech*'s editor. He clarified that the group consisted of students, religious groups, and labor organizations working to roll back racial discrimination. He explained that they promoted nonviolent action. He walked readers through CORE's action policy, the official playbook MIT-EPIC followed. He justified singling out Woolworth's. But most important, he called attention to the truth as he saw it, and the need to press forward.

Racial discrimination is not a regional problem concerning only people who live in certain areas. Racial discrimination everywhere concerns everyone. It is your duty as a citizen of a free country to oppose it actively. MIT-EPIC wished to commend The Tech for their excellent job of encouraging civic interest in the MIT community.[17]

The spring 1960 semester began to pass. Classes gave way to final exams, graduations, and summer. School let out, but Harvey Pressman announced that EPIC was open for business through the summer. Still, the throngs of students making their exodus from the Boston area for the break prompted calls from the NAACP.

Approximately 100 former picketers are still on the job at the Woolworth stores in Roxbury, downtown Boston, Cambridge, Mattapan, Belmont, Lexington, Quincy, and Lynn. But the lines are small....Many of our members have been picketing faithfully since March, along with other members of EPIC....But this summer we must do more![18]

Meanwhile, EPIC's relationship with CORE showed signs of strain. A terse conversation between leaders of the two groups began late that April.

Dear Mr. Robinson,

This is largely a confirmation of our telephone conversation. We have enough leaflets to last us until but not including May 7th. We intend to continue picketing through the summer. We would like, if possible, 100,000 leaflets shipped the cheapest way possible. If the cheapest way won't get them to us by May 7, then a couple of boxes might be sent by a more rapid method and the remainder shipped in less expensive and more leisurely fashion. Although we prefer the EPIC imprint on the bottom, it is not essential and certainly not worth running already printed leaflets through the press again. However, we would appreciate it on a new printing.... EPIC has set some money raising activities in motion and I very much hope we will become self supporting after this shipment of leaflets. In the meantime, we very much appreciate your support.[19]

William Gamson had written this letter to CORE on EPIC's behalf. He was trying to help, but his request came across as a bit pretentious. Unreasonable even. Robinson responded a week later.

Dear Mr. Gamson,

We are shipping Trailways Bus Express <u>collect</u>[20] to you at the above address 15,000 of our CORE Woolworth leaflets without imprint.... If you want another 100,000 with the EPIC imprint, please send the EPIC imprint with any required changes. We can then consider whether to go ahead with them.

It was our clear understanding that we were to receive a contribution from EPIC after the fund-raising concert. I should like to know the details of how that arrangement was changed—was the concert arranged and then the fund provision changed? If so, it would seem entirely possible that your organization could get some of the proceeds back. CORE has increased its field service, has stimulated the picketing and boycott in many areas of the country—and in so doing has spent many thousands of dollars. If we are to continue, we must have help. I know you understand this and trust that your group will see to it that funds in the future are fairly distributed.

Robinson referred to the fact that calypso singer and civil rights activist Harry Belafonte had headlined a sold-out concert, organized by EPIC the week prior to Gamson's initial letter.[21] Robinson ended his letter, signing off as CORE's executive secretary. He added a postscript. He told Gamson that Trailways did not deliver packages. The entitled asshole would have to get off his butt, go to the bus station, and pick up the leaflets himself![22]

Harvey Pressman took over all correspondence after that. He and Robinson conversed over the phone and through the mail, mostly about exchanging picketing materials.

I'm sorry about the misunderstanding about money. The truth is that we're both honest and in debt; two conditions that frequently occur simultaneously.

Pressman's letter to Robinson squared financial arrangements and secured future cooperation. But their correspondence also revealed that EPIC wanted autonomy. As it did, the group sought new allies to fight discrimination in the South.

The timing was fortuitous. Two potential suitors emerged. One was Students for a Democratic Society (SDS). That group formed on May 16–17, 1960, when a group of students eager to join the civil rights struggle gathered at the University of Michigan. Partnering with SDS would have offered EPIC an opportunity to scale its ranks and expand its focus. But EPIC declined SDS's overtures to join forces.

But another opportunity soon came along that offered EPIC a chance to intensify the work they'd already begun. On April 12, 1960, Curtis B. Gans, the national affairs vice president of the Philadelphia, Pennsylvania–based United States National Student Association, had sent word to Ms. Ella

Baker. Baker was an official with the Southern Christian Leadership Conference.

Dear Mrs. Baker:

I look forward to attending the Southern Leadership Conference meeting this weekend at Shaw University bringing together leaders of the southern student movement from all over the South.[23] *The National Student Association, as you know, has taken an exceedingly strong interest in the southern student movement and we are very much concerned as to its future.*

On the weekend of April 16–17, 1960, students from across the country converged in Raleigh, North Carolina. Their destination was Shaw University. There they discussed, debated, and proposed action steps to begin a student-led movement of their own. They aimed to fight racial discrimination in public accommodations. On the first day of the meeting, Mr. John Cooley reported that *working group #10* had drafted and affirmed a statement of purpose. First and foremost, they solidified their commitment to nonviolence. The working group also recognized their responsibility to help sympathetic northern groups. Their priorities were twofold: develop an effective communication strategy for their respective constituents and raise and distribute money for legal aid.

The proposals—recommended and ratified by the assembled body of students—birthed the Student Nonviolent Coordinating Committee (SNCC).

SNCC convened its first organizational meeting as planned on May 13–14 in Atlanta, Georgia. There, they made more plans. Notably, they plotted to expand and strengthen their interracial ally network throughout the South and North. On June 14, 1960, Jane Stembridge, SNCC's secretary, sent a form letter addressed to EPIC. SNCC had sent the same letter to a number of groups it hoped to recruit to join its fight.

Dear Sirs:

The Student Nonviolent Coordinating Committee was established at a meeting in Raleigh, N.C. on April 17, 1960.... While our Committee is self-directing and not organizationally related to any other groups working in the field of human rights, we are writing various organizations to send observers to our meetings.... We would be happy to have EPIC send observers to our meetings.... The observers are allowed to attend all the meeting and may be called on to help the group in discussions where their expert knowledge would be of benefit.[24]

Pressman declined to participate. He explained his reasons in a short, handwritten letter.

Dear Ms. Stembridge,

Please excuse me for not replying to your June 14 letter. It was misplaced in the process of moving. We regret that we are physically and financially unable to send an observer to your meeting. However, we would very much appreciate receiving copies of the minutes of your monthly meetings.

Very Truly Yours,

Harvey Pressman, Chairman, Emergency Public Integration Committee

4 Humboldt St., Cambridge, 38, Mass.

With its delayed writing, brevity, and message SNCC could have taken Pressman's letter as a simple dismissal. But they did not. They continued to push Pressman for any current and future help EPIC might offer.

Dear Mr. Pressman:

Thank you for your letter of August 8. We are extremely sorry that EPIC cannot, at this time, send observers to our monthly Coordinating Committee meetings. However, we understand your situation. Perhaps in the future you can join us for some of the sessions. The invitation remains open. We will be delighted to have you.

Regarding your request for copies of monthly minutes, we will send what material we publish, probably contained in the newsletter "The Student Voice." It is hoped that this newsletter will be issued shortly after each meeting.

Likewise, we want to receive news releases from EPIC. Any assistance which your organization can give to us will be deeply appreciated. Best wishes for the future.

Sincerely yours,

Marion S. Barry, Jr., Chairman

But Chairman Barry did not have the insight to read between Pressman's lines of rejection. The truth was, EPIC wasn't what it was just a few months before.

Scales of apathy began to wither from northern students' eyes. EPIC's relationship with CORE and SNCC deepened. It continued to join the fight against discrimination from Boston down to Mississippi. But something was missing. In fact, many people were missing. Individual students' voices faintly echoed through the discussions and actions that continued through 1961. But MIT's voice fell presciently silent. Through another academic year, Boston EPIC and other students in northern communities were intensely engaged, but the student struggle against discrimination never again appeared on, let alone topped, MIT's student agenda. The student body continued to make token efforts to engage. They invited Malcolm X to speak.[25] They welcomed James Baldwin.[26] A few other civil rights leaders appeared—all before disappointingly small crowds in MIT's Kresge Auditorium.

When MIT-EPIC students had first started picketing Woolworth's in 1960, James Geiser, a New York City native, spearheaded their organizational efforts. He said that EPIC *was the beginning of a new epoch at MIT, the transition from an apolitical school to a political one.*[27] Geiser's prognostication would never come to pass.

Students had come to MIT to be transformed into what the institute promised to make them: leaders, capable of, and charged with, fashioning better tools and machines to support and extend America's political and economic standing in the world. They were building the technologies and machines that would soon begin to govern the nation. And they did it with little care or concern for black people, either within the greater Boston area or throughout the United States. This epic lack of concern would make it palatable, easier for MIT to lend its knowledge, its computing resources, its computational innovations, and its ties to industry and government—a government that began to see black America as an enemy target for the country's new computing machines.

* * *

By 1971, Mel King had joined MIT as an adjunct professor in the Department of Urban Studies and Planning. There, King—who began his career as a math teacher—inaugurated the Community Fellows Program. MIT spotlighted the new program, and King's role, in its 1971 Annual Report to the MIT Corporation and broader institute community.

The Department of Urban Studies and Planning has established the Whitney Young Community Fellows Program to help minority leaders cope with the social and economic development of their communities. Developed by Melvin King of the New Urban League and by Professor Rodwin, the main aim of the program is to enable a selected group of local leaders to spend the equivalent of an academic year at M.I.T. working with faculty on projects of special importance to the Fellows and their organizations or communities.

Community Fellows will be drawn from the staffs and departments of newly-elected black mayors, minority community development corporations, model cities and poverty program staffs, private minority organizations with action and development components, legal assistance and information offices, and other minority enterprises. A third to a half of the Fellows are expected to come from the Boston region.[28]

In his MIT capacity, and later through his position as state representative, King served the black Boston community in many ways. He helped to raise technology's profile within the community, and he helped to mediate technology's impact on it. In 1983, King, an unlikely prospect, decided to run for mayor. And William was about to make a chance encounter.

His South End campaign HQ was three blocks from my front door. His son Mike had introduced me to a community college student who I began to do tech calls with. I had an Osborne I, the "world's first portable computer" made by Adam Osborne. I popped into the campaign office one day to see what was up. As an activist with The Black Community Information Center (BCIC), a Roxbury Malcom X–based nonprofit, I volunteered there. Mel's work at MIT Community Fellows program intersected—he attended BCIC meetings. I taught people how to fix their own computers. It was a community-empowerment-type organization.

The day I went in, his campaign manager said we need more computers to database volunteers. I lent my OS and later replaced it with Radio Shack–donated TRS 80s. I ended up being the campaign manager's IT guy. Did whatever they asked. Judy managed the office. Mel's house was around the corner. My mom visited during the campaign, read Mel's Chain of Change book and liked it. She was NAACP. Jesse Jackson was running for president under Rainbow Coalition. Mel adopted it, but I think he leaned to [the] Green Party. I joined the Green Party. Worked at Mel's HQ doing odd stuff and was the processor (using Osborne and Supercalc Spreadsheet) of the real-time vote tallies from City Hall so Pat Walker, his operations manager and math genius, could call it and manage field placements. Mike King was in charge of Roxbury Community College's technology program.

Mel King made history. He won the mayoral primary campaign, garnering almost 99 percent of Boston's black vote. But he lost the general election to Ray Flynn—his opposite in almost every way. That campaign crystallized both King's and William's work. Both would connect technology to the black community long into the future. Just a short time later, King went on to found the South End Technology Center. The center was an outgrowth of his work at MIT and aimed to boost technology use in Boston's black community.

As for William, 1983 jump-started much more for him than just campaign work.

It was luck! I was an electronics instructor at a science tech magnet high school in East Boston.

Boston's Mario Umana Harbor School of Science and Technology was established in 1975, in partnership with MIT. It was part of Phase II of Boston's court-ordered school desegregation plan.[29]

I got parts and stuff to use in class—they had no budget. I was sent there—by Wang Labs where I was a VS100 field engineer—to help out the public school. Guy before me just walked out. There was a kid I mentored named Vishal DuBey, his dad was an IBMer, built a breadboarded microprocessor from discrete components. We entered it into the Boston Science Fair. He won it! First time this school had won in eight years. The Globe ran a story on it.[30]

While scrounging parts from MicroTech, a Commodore dealer, they hired me to fix stuff. Summer came and fixing became full time. The Commodore 64 had hit and it was a big seller. So Tom Sinopoli, owner of MicroTech, said if I took all these Commodore parts over he would allow me to own the service shop. Tom's Italian. Microtech was in the North End. Tom wanted to get out of the business. I took in the debt—feeling like we could make it back, and it worked out. That takeover put me $6,000 in the hole. But everyone gave us Commodore law office service contracts as these other dealers began to establish IBM PC dealerships.

I did not want to go back to that school! I converted the name of the company to MetroServe during my first summer off teaching. Lucked out. The Commodore 64 had a parts shortage. But Alan, a black kid from Roxbury Community College I hired, figured out how to fix it with Radio Shack parts. So Toys "R" Us sent over one hundred failed units. I got paid a flat rate to fix them all, plus getting the Commodore Business Machines models in to fix from big businesses downtown made it lucrative.

When MetroServe hit, it was my second business attempt. Even at IBM and Duke Power, I sold Vanguard fire alarms on the side and tried a number of get-rich-quick schemes. "Work for self" was the mantra. I thank Black Enterprise and Ebony *biography reads for that influence. My dad retired as a postman. But on the side he was importing jewelry from war buddies in Belgium, and selling that on his mail carrier routes. He advertised in* Black Enterprise....

Now in business with a silent partner, William's MetroServe Computer Company generated $7,500 in revenue by the end of 1984. By the end of 1985, the business had raked in $185,000. When William finally incorporated on September 10, 1986, he was pulling in close to $500,000. And, he had relocated his business across the Charles River, to Cambridge. By the early 1990s, William owned Boston's largest, and Cambridge's only, black-owned computer store.

I always felt that I was confined to certain boxes, that my ultimate advancement from those boxes was controlled by someone else.

Not anymore. William was living his dream. He finally got to do his own thing. And he was not alone. But that was hard for William to see at the time. The 1980s had created a stark divide that cut right through the nation's old racial politics, and the growing computer revolution.

Figure 3.1. William Murrell, president of MetroServe Computer Corporation, 1990. Courtesy of William Murrell.

Chapter Four

THE VANGUARD

William and Derrick were not anomalies. In the late 1970s, and in earnest by the 1980s, a torrent of black technophiles began making moves. They shifted the relationship that had made their parents and grandparents technology's victims. And they propped themselves up to be more than just spectators in the coming computer revolution.

They were the Vanguard. They were different people, with different preparation, pursuing different interests. They would discover one another soon enough but at that time they were each on their own journeys.

They did not know they were leading a revolution.

Lee, LA's Voice

It may sound corny, but I just believed I was born to do it, and came to the point where I did do it. I moved on those feelings and became a DJ.

Lee felt attuned to radio's power, practically since birth. That also happened to be about the time that radio became popular. Born in 1948, Lee Bailey would become the the Vanguard's elder statesman, in more ways than one. Lee spent only a few years in Moreland, Georgia, before heading to Pittsburgh, Pennsylvania. There, he cultivated his fascination with radio. He made the studio at WZUM radio his second home. The DJs at the station let him hang around. For the privilege, Lee ran errands and performed odd jobs. In 1966, when it came time to leave his electronic cocoon, Lee joined the Air Force. But instead of reenlisting in 1970, he went looking for his first love. Like trailing a pied piper, Lee followed radio voices, *town to town, up and down the dial*,[1] from Flint, Michigan, to Sacramento, California; Washington, DC, and Los Angeles.

From 1975 on, Lee made Los Angeles his home. And he immediately began honing his trade. He found work at K-DAY radio. Then KJFJ-AM. Finally, KUTE-FM. There in that KUTE DJ booth, Lee discovered the news. Journalist Steven Ivory helped him make the connection between music—Lee's usual traffic—and public affairs. Ivory was a columnist for *Soul Magazine*. People in the business, and on the street, in the 1970s called it the black *Rolling Stone*.

Lee recognized something revolutionary while he dabbled. Radio may not electrify every listener. It may not move everyone to lift their voices on air as it had Lee. But Lee witnessed it. The content that filled radio's airtime pauses connected people to those on-air voices. Musicians. Entertainers. Talk show hosts. And yes, even the folks that brought the news. Even the folks who colorfully dramatized advertising spots to generate station revenue. Apparently, none of Lee's LA market colleagues shared his epiphany.

When I was on the air as a jock—it was me and Steve Ivory. And he would call up and we would talk about the music that was popular at the time—we'd do a little five-minute stretch in my shows in the evening. I was on the air from eight P.M. until midnight. And so somewhere between nine and ten we'd do this little segment called "the Music Man." As a result of that, people were curious about music personalities they were hearing on the station. If anyone puts a song out, there were a thousand curiosities about the new music and person. There were pop magazines and soul magazines with information about R&B celebrities. But there were really no radio programs, no information programs about urban entertainment. I got the idea—it became more and more obvious—that people wanted that information from the radio. Nobody was doing that. Now, you had countdown shows. . . . But you didn't get the scuzz, the inside info, as it were. You didn't really get the news behind the entertainment industry.

Lee aimed to be LA's voice. He gave the people what they wanted. This was Hollywood, after all! All people—including, and especially, black people—wanted to stay connected to the women and men whose voices tickled their ears in the morning. The voices that made them tap their feet during an otherwise excruciating commute. Listeners wanted to stay connected to the words coming out of that radio. Lee recognized his influence. His power. More important, he recognized that LA provided a ready-made market.

By the early 1980s, Lee started spending more time in his garage studio producing radio ads than manning his DJ booth. Then, he and his wife, Diane Blackmon Bailey, formed *RadioScope* in 1983. The show gave people the inside scoop about the celebrities, entertainers, and other personalities with whom listeners had developed a deep relationship.

I remember one of our affiliates here in Los Angeles—K-ACE. The program director, Alonzo Miller—he had written Superfreak *for Rick James. He was the program director of the station, and he said he was going to do me a favor. He had agreed to run our five-minute program, which was daily, Monday through Friday, and an hour-long weekend program. He said, I'm going to do you a favor and only run it on Monday, Wednesday, and Friday. 'Cuz, I just don't see how you're gonna have enough information for everyday. I said—please! Don't. Do me. Any. Favors. Please!*

But I could see his view. He thought having a five-minute program on a daily basis with enough information was just unimaginable. I mean you had Entertainment Tonight *but that just didn't carry over.* RadioScope *was kind of modeled off of* Entertainment Tonight. *But* 60 Minutes *was more what I modeled the show on. Just the fact that a program director in a major market did not believe there would be enough information for a daily radio program, an information program, was just astounding.*

Infotainment was the cornerstone of RadioScope's content philosophy. *Information that's entertaining, and entertainment that's informative,* as Lee put it.

That covered a multitude of sins. We were never restricted to having to do just music, or whatever. It's not just black folk. It's everyone. They're drawn to entertainment. They're drawn to celebrities. So you bait 'em with that then give them something else. Doesn't mean they're necessarily going to focus on the news that they ought to, but at least you're putting it out there. It's like putting out food for a pet—the craziest analogy—but you put it out there, and hopefully they'll eat it. And maybe you mix in some vitamins or nutrients—things that will make 'em healthy—and they'll get it, without even knowing they're taking it.

Lee's nationally syndicated *RadioScope* brought the right philosophy to radio, and the program quickly found its audience. But advertising powers radio, and for the first few months, Lee struggled.

The first couple of years, there was zero coming in. It was just through the grace of God. I have no idea how we did it. 'Cuz I didn't have any money. I guess the idea was big enough that it just held on. Somehow, some way, it held on. At about two or three years in we attracted the attention of a great man, as far as I'm concerned—Chuck Morrison, from Coca-Cola. He heard the program one day, and thought it would make a great marketing vehicle for Coca-Cola. He used to be in charge of urban and Hispanic marketing for Coca-Cola. He heard the program and he got in touch with us. And that's how things took off. [2]

It also didn't hurt that *RadioScope* gave America its first taste of Prince's *Purple Rain* before it debuted in 1984.

By the start of the 1990s, people recognized Lee and his recently incorporated Bailey Broadcasting for more than just spinning records. His reputation towered high above *RadioScope*. His value exceeded his marketing prowess. He grew larger than the ad campaigns he developed around iconic black popular culture productions like the one he did for the film *New Jack City*, a New York City–based film about the crack trade.

Lee used Langston Hughes's introduction to *Raisin in the Sun* to open his 1993 *Los Angeles Times* profiled documentary *Up from the Ashes: The Los Angeles Rebellion and Beyond.* Lee's radio documentary resounded black voices—voices that American audiences needed to hear. Voices that America needed to fully wrap their minds around when seeing Rodney King brutally beaten by LAPD cops. The *Times* described how Lee fulfilled his own broadcasting philosophy through this groundbreaking radio event.

Up from the Ashes *was neither the first nor the only recapitulation of the rioting. But it was unusual because it dealt specifically with the emotions, opinions, and history of the African American community. The show originated from Bailey Broadcasting Services in the San Fernando Valley, one of the independent producers scattered across the country that operate outside of radio's predominantly Caucasian mainstream. "We make no bones about it—our programs have a black edge," said Lee Bailey."…'If you're black, being black is on your mind 24 hours a day. There are always issues being rais'd.'"*

The newest transformation in electronic media hit the world about the same time that Rodney King's internationally circulated video dropped. The transformation would spark a new media era for black people, and for content distribution that pushed our political interests. Lee was as well positioned to drive this moment as any of his soon-to-be digital compatriots.

Figure 4.1. Lee Bailey, Los Angeles DJ and later owner of the Electronic Urban Report (eurweb.com).

TYRONNE, THE SYSOP

I was a working musician and I was in my fourth year in college…

But Tyronne was hooked on his music. And it brought him to a crossroads. He dreamed of becoming a classical saxophone player. Where he got that idea, Tyronne never really knew. He was, admittedly, a little nerdy. He learned music theory and played a horn in front of music teachers and other students. But that did not provide the same thrill that Norfolk's nightclub scene offered. The adulation. The instant gratification. The women!

I played around these clubs in Norfolk, VA. That's where was I born and raised.

What happened was, I played and enjoyed it. . . . Played everything. . . . When I was in middle school, I got lucky. The guy who was my junior high school band director was the band director at Norfolk State University. He offered me a Full.

Boat. Scholarship. And I knew the University of Michigan was too cold! And I did not want to go! I was offered a partial scholarship there, too.

But what happened was . . . I played, ya know. Enjoyed it. I played everything. You name it, I played it. Named it. Sung it. Sang it. You know—I did it all.

Tyronne convinced himself that the chance to complete his degree had an endless shelf life. So he told himself he was going to have his fun.

I told myself that when I get thirty years old, I was going to get a regular job— regardless of what it was—not for the money, but for the benefits.

Tyronne dropped out to play his music. But things didn't play out quite as he had expected. By 1976, he had met a woman. They got married. The promise he had made to himself (and her) sat in the back of his mind like a metastasizing tumor. He worried about finding that job more than completing his degree. He needed a good one, and with benefits.

So I got what I thought was my dream job. It was with the government. It was a great job because it had good benefits. I wasn't worried about the money. I wanted the benefits.

Tyronne soon got another job opportunity that took him to Richmond, Virginia. It didn't last long. But another opportunity quickly came his way.

The next place was called Systems Engineering Computer Company. They needed someone to take night computer transmissions from a cassette machine called a SYCOR. [A SYCOR was a data-processing device used to transfer and process data remotely.]

You take it, you copy it to a mag tape. . . . I- I- I- I didn't know a thing about it. But I did it, ya know. And then the company began selling second-generation, IBM PCs.

Well, my wife at the time was doing legal term papers for the University of Richmond. Their students. So I bought her a PC. Somebody gave me a game. I put it in there. It said I needed a color graphics adapter card. I went back to my supervisor, who happened to be the field engineer at the time.

Hey. How much would it take to fix this, to bring it up to snuff?
      Oh—I can do it.
      No problem!

He was gone for fifteen minutes. Came back, and he gave me the bill. I looked on it, and it said—$140.00.

What did you do for $140.00?
    Well…I- I- I had to replace the card.
    What did you physically do?
    Um…Five screws.
    What? What? What? What? What?
    Took the board off. Took my board off. Then another screw. Put another board in there with my screw. Hit a coupla dipswitches then put it back in.…
    You charged me $140.00 for that?
    Yeah.

Then I decided to read the books that came with that thing. I continued, and they promoted me to a computer technician. Then a job control language technician.

A job control language describes jobs for batch processing, specifically on the IBM S-390 mainframe computer.

Then, the people who repaired my wife's computer [were] Digital Equipment Corporation. Next thing I know, they needed somebody … 'cause PCs were really getting hot. And I knew how to build 'em. Knew how to repair 'em. Well, my company, Systems Engineering, they opened up a sideline company called PC Consignment Center. We'd sell used computers. Canoes. Fishing equipment. So I said, let me start a BBS, [bulletin board system] ya know, for the business. Then DEC came around and offered me a job as a field engineer. Because PCs were coming up. And DEC, they were stuck on stupid.[3] So I did. But I decided to continue the BBS at home, so I could learn computer networking. Computer communications. And a little programming.

Back in the 1960s, MIT had started a new computer networking project called Project MAC. Those early computer scientists discovered that *when more than one person can use a computer at the same time, those people will use the computer to communicate with each other.*[4] That was certainly true for Tyronne. The potential to send, receive, share information, or communicate in real time with other people—that is what motivated him to invest thousands of dollars over the coming years building his own BBS.

* * *

Several technological developments paved the path that Tyronne and the Vanguard ventured down over the next fifteen years or so. Each operated, more or less, concurrently throughout the 1980s. ARPANET. BBS. FidoNet. Usenet.

The Department of Defense, collaborating with several universities and labs in 1969, built the ARPANET. It was, simply, the first computer network that utilized a specific instructional protocol. That protocol directed how computers could talk to one another. They named the protocol TCP/IP, or Transmission Control Protocol/Internet Protocol. This model and its operation launched the Internet. From 1969 to 1981, a relative handful of government, university, and private scientific lab personnel accessed and controlled ARPANET.

In 1979, Ward Christiensen and Randy Seuss wanted to communicate with other members of their Chicago computer club. They built a computer bulletin board system (BBS). This was an open network that connected personal computers. Users dialed one another's system using a telephone line. Once connected, users posted and exchanged content, text-based messages, software, and information, like credit card account numbers.

In 1984, Tom Jennings created "FidoNet." FidoNet included the same basic BBS functionality. But it distributed information differently from the original BBS. FidoNet utilized the same kind of store-and-forward technology that had powered IBM's Personal Office System in the late 1970s. This was the same system IBM propelled to popularity in the early 1980s. This kind of system allowed individual computers (or a group of several computers) to serve as a host or hub computer. These hub systems stored information and then they automatically passed on the information to other computers connected to the hub.

System operators (SYSOPS, for short) owned and operated the hub systems. They assumed the costs of purchasing the necessary hardware and software to run their system. They also paid the phone charges needed to regularly connect to the network. SYSOPS maintained a roster of both unique and shared users. That is, a user might be connected to one or more BBS. Doing so connected them to each SYSOPS network. This was an Internet. And SYSOPS served as de facto Internet service providers (Internet Service Providers).

A few years before FidoNet, in 1980, a pair of Duke University graduate students developed a hybrid system. Tom Truscott and Jim Ellis built what they called Usenet. It stood simply for "Users Network." Usenet shared characteristics with ARPANET, BBS, and FidoNet. Usenet newsgroups, like those of a BBS, allowed users to post messages (news) on a system. In the early years, users posted mostly technical information. As time went on, people posted more socially oriented content, in theme-based groups. ISPs created new groups, based on users' interests, and governed by their own protocols.

Usenet operated on a distributed system. With both BBS and FidoNet systems, information was routed through individual hub systems. If I were a SYSOP one day, and got rid of my system the next, users on my network could no longer connect to me or anyone else in my network. Usenet trafficked data through redundant systems, operated by larger ISPs. Universities typically served as ISPs. And they almost always connected to the ARPANET. That once-closed network began to open up in 1985 when the government authorized NSF-Net, the National Science Foundation Network.

Because pieces of information were packaged into files and distributed across many computers in the network, that information remained accessible, even if any one computer disconnected from the network. Most large ISPs connected to ARPANET. So Usenet directed greater volumes of data traffic through the system more rapidly than any prior system.

* * *

In 1987, Tyronne started a BBS from his home in Richmond, Virginia. He named it InterCity BBS.

I wasn't into networkin', or anything like that. But I had a friend who was another system operator. He was into geneaolgy. He asked me if I could pick up the National Genealogical Society Newsgroup. I didn't know what a newsgroup was.

Mix Master Ken

Since the 1960s, American presidents, New York City mayors, corporate CEOs, and civil rights icons have made the Brooklyn neighborhood of Bedford-Stuyvesant their pet project. It was a ghetto in the 1980s. Almost all black. It depended on the block, of course. But anywhere from 16 to 40 percent of Bed-Stuy families lived below the poverty line. With little exception, only about 4 percent of its adults graduated college. In most areas, scarcely more than half completed high school. Unemployment ranged from 3 percent in some areas to more than 7 percent in many—even to 12 to 18 percent in some sections.

Bed-Stuy, like most US ghettos, provided very few on-ramps to America's prosperity highway. But Ken Granderson found one. That on-ramp started at his home at 349 Stuyvesant Avenue, and it traveled four blocks south to

Fulton Street. It then snaked west for a four-mile stretch. When Ken started his journey, the neighborhood's complexion was more than 90 percent black. When he crossed Washington Avenue, it lightened to just 65. When he passed Hoyt Street, only two of every ten people looked like him. And by the time he reached the end, his world was as white as an Augusta, Georgia, country club. And it was densely packed with networks of opportunity.

From third grade to twelfth grade I went to St. Ann's school in Brooklyn Heights.

School was mostly white. Out of a graduating class of seventy there were ten black students, six female, four male.

None of those students happened to be in Ken's computer class—the one that had first lit that spark. The one that showed him that computers and new technology were not something to fear. In fact, that class taught him quite the opposite.

Somewhere either sophomore or junior year or so, the school got a computer, a DEC-PDP-11.

This "minicomputer" by Digital Electric was one of its most popular machines on the still-nascent personal computer market. Among other advantages, the machine provided peripheral ports. This meant you could input and output data into and from the machine, from and to other sources and destinations. This feature alone had single-handedly imprinted the power of computer programming onto Ken's consciousness. It impressed upon him the idea that software is flexible, malleable, controllable, meant to fulfill any whim and fantasies its developer wished to chase.

From that computer class at St. Ann's, Ken secured another opportunity.

I got to take classes at Brooklyn Polytechnic on Jay Street, and so we got exposure to a couple of programming languages. We did some Fortran with the punch cards. But the big thing was the fact that we had computers with terminals.

Ken added another language to his programming toolkit when he was invited to take part in a summer program across the East River, in Manhattan's storied Greenwich Village neighborhood.

New York University's Courant Institute for Mathematics was a hub for what was then becoming the field of computer science. The institute—particularly its location at Warren Weaver Hall on the corner of Mercer Street and West 4th street—was also symbolic. In 1970, the building housed a new, state-of-the-art, $2 million computer.

Richard Nixon had just ordered the invasion of Cambodia, raising the ire of anti-war activists across the United States. And the week before that, police had killed four protestors at Kent State University. On May 5, a group of 150 protestors stormed Warren Weaver Hall and "liberated" the computer center. Then they threated to detonate a homemade bomb that they had strapped to the towering computer console. To the protestors, that computer symbolized and showcased state power. They believed the state used that computer to exercise control and advance its own interests and prerogatives. To those protestors, that computer extended the oppressor's reach.

Having seized its oppressors (the computer and its operators), the protestors demanded that the university pay a $100,000 ransom. The protestors planned to use the money to bail out Black Panther Party members indicted for conspiracy to bomb department stores and police stations, and murder policemen.[5] Police ultimately thwarted the protestors' intentions and heroically rescued the computer!

Almost a decade later, in 1979, Ken stood in Warren Weaver Hall, not to liberate that computer, but to learn how to use it. He was participating in a summer program that introduced high school students to computer programming. Their curriculum focused almost entirely on learning Artspeak. Artspeak was a new computer language, developed in 1970 by Courant Institute professor Jacob Schwartz. Artspeak was a grid and coordinate-based program. Users transferred their program onto a sheet of paper before feeding it into the computer. Take this simple program, for example.

```
LET P5 BE POINT (5,5)
LET P1 BE POINT (5,2)
COPYPOINT P1 TO P2
L1 ROTATE P1 ABOUT P5, ANGLE 5
LET C1 BE LINE P1, P2
DRAW C1
L2 ROTATE P2 ABOUT P5, ANGLE 2.5
REPEAT L1 TO L2, 143 TIMES[6]
```

These simple lines of code produced an exquisite line drawing, an image appearing as if multiple, winding interlocking staircases were drawing your eyes down to some abyss.

The program was relatively simple, but Ken and his friends grasped enough key programming skills to have a field day back at school.

We basically pirated the source code of a Star Trek program that printed out on teletype machines that you would sit in front of. They were about desk-high, you know. It would just print out—every time you moved, it would print out a new little map of your quadrant and space, and every action you took, like a keystroke. It would print out a new map of space and where the ships were. And we printed out the whole, basically source code of that, and converted that to the PDP-11 version of BASIC, and used what they called Direct Cursor Control, which, even though it was a direct computer screen, you had the ability to, you know, to print characters on particular coordinates on the screen. So we converted it from a game where you would have to wait for the entire quadrant of space to print out for each time, to one where when you took an action, you could actually see the phaser dot move across the screen and hit the Klingons.

Artspeak didn't survive long beyond the summer of '79. But it left an impression on Ken, one more significant than any computer game. Ken spent every day of his young life with one foot in what some saw as an impoverished slum. The other stood in a white, wealthy garden teeming with opportunity and connection. Ken recognized that both contained tremendous value. He would soon realize, and he would soon demonstrate to others, that software had the power to bring both his worlds together. For the time being though, he chose to master something newly produced from his own streets.

St. Ann's. Brooklyn Polytechnic. NYU . . .

That's where I got the bug. But because my family didn't have the finances to get any sort of computer, I ended up getting into DJing instead, and didn't really do much with computers.

When I came up to Boston to go to MIT in 1980, I was thinking of—I first was interested in computer science. But the level of nerdism was so intense, that, as a seventeen-year-old moving away from home, and someone who was a nerd, and wasn't that socially adept, I said the last thing I wanted to do was hang out with the dork crowd, The Dungeons and Dragons crowd. So I ended up not taking computer science, but doing electrical engineering when I was in college. Took a few computing courses, and then rediscovered my passion for programming in the very late 80s, early 90s.

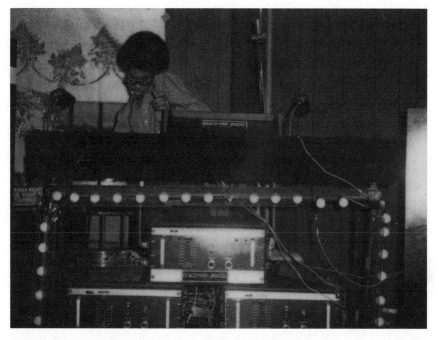

Figure 4.2. Ken Granderson, nurturing his DJ hobby as a teenager. Courtesy of Ken Granderson.

THE MONEY MEN, DAVID AND MALCOLM

David Ellington had to get from zero to a million dollars now. That's how he approached life after high school graduation. He had designed his own future, down to the most minute detail. What education to get. Where to get it. What to study. Who to meet while doing it. How to get from preparation to launch point as soon as possible. David was anything but modest. He exuded confidence, and was in complete control of his future.

Bottom line is, I was living in Tokyo from '85 to '86 teaching English—just conversational English—and made friends there. It has a very small black community, so we all hung out together. I moved back, went to Georgetown Law School in DC.

David had already earned a BA in history from Adelphi University in New York and an MA in comparative politics from Howard University, one of the country's premier Historically Black Colleges and Universities (HBCU). He concentrated in international and tax law at Georgetown.

Then I actually went up to Cornell and studied formal Japanese after I graduated Georgetown to get a certificate in Japanese, to get my second language.

So I'm now back in the States [in the late 1980s]. *I decided I wanted to be on the Pacific Rim instead of being on the East Coast, because I wanted to do international trade work, focused on Asia, specifically Japan, and use my Japanese. And so I was doing that. Attempting to do that. Trying to get a job in LA in 1990. And basically that's when the first wave of a recession hit Japan. So all the Japanese-related work was drying up for law firms in LA. There's a huge Japanese community in LA.*

David moved to Los Angeles. He shifted his legal practice and pivoted toward his new professional course with ease.

Malcolm was ten years David's junior. He had grown up on a farm in rural Pennsylvania.

Very rural. We had like fifty . . . sixty acres of land. Animals. Lots of fields. When I was a kid, I got exposed to computers at school. I begged and begged to get one at home. I played with it, and of course I lived on the ones at school. And I even had teachers who would let me borrow them on the weekends. So I got to play with them all weekend. And then eventually, I was able to get one. I was a teenager, ya know, fourteen, fifteen years old. . . . That was the beginning. I knew early on that I was interested in math and science, but I didn't know where I wanted to go with it. [Computers] *were just my hobby, ya know.*

The early 1980s brought the personal computer and new adventures in computer networking even out to the farm. BBSs. Usenet. Filesharing. These became Malcolm's playground.

That's all I did. I was constantly annoyed when my parents would pick up the phone and interrupt my slow BBS access, because I was using it. We only had one phone line, so if a call came in, or they picked up the phone—that was a serious interruption. But I was on a lot of bulletin boards. I wrote software. I downloaded software. I hacked games. I had a constant thirst, so I was just constantly reading information, downloading information. There were manuals that you could download from bulletin boards that could explain more and more, and that was, sort of, my teacher. The forums—asking questions, getting answers. . . .

Malcolm found great teachers in the bits and bytes distributed across the 1980s Internet. They were far better equipped to quench his thirst for technology and entertain his intellectual curiosities, which ran ahead of most of his peers—and, likely, many of his high school teachers. He left them all behind a year early when he headed off to MIT in 1987.

It's funny. I started out in materials science, because I was very curious. And then I had an advisor who was an African American, who oversaw the history of science and technology. The one thing I'll never forget he said. He said I was taking the same computer science course that all of my classmates were taking, but I was the only one not complaining about it. I was loving it, and I didn't think it was that hard. So he said, "I think that should be your major."

I was very curious about how things work. I was training on electron microscopes, and I was making superconductors. I was using lasers to melt objects to make superconductors, making high-tensile strength conductors for DoD experiments. All of this at sixteen years old. It was really fun and exciting! So it was hard to give up being in a lab, and melting things and looking at things at super-high magnification. But eventually I chose computer science.

Malcolm took advantage of every opportunity. He wedded himself to every computing challenge that MIT had to offer him. Perhaps none was more exhilarating than his work with Project Athena. MIT's president began the institute's 1983 annual report announcing the expansive and ambitious project. MIT designed it to capitalize on the personal computers' advent and rapid adoption.

An example of the kind of vision and action so characteristic of our faculty and academic leaders is the inauguration this year of a massive educational experiment aimed at exploring how computers can accelerate and enhance the learning process. Started with resources provided by the Digital Equipment Corporation and IBM, the project is based on the premise that the next generation of personal computers represents a revolutionary new medium for learning. . . . Supporting the experiment will be thousands of personal computers and terminals, reinforced by mainframe computers, storage devices, and printers—all integrated into a single system. In addition, both IBM and DEC will have at least five representatives stationed at MIT to help develop the necessary computer networks for this project and to work with students and faculty in blending computers into the educational process. . . .

The ultimate aims of this joint venture are to find out exactly how these powerful tools can aid students in visualizing difficult concepts, breathe new life into laboratory experiments, help develop the skills, knowledge, and insights needed for design problems, and even help cultivate that elusive talent we call intuition. This work has the potential for helping universities educate students for many disciplines—engineering, science, architecture, humanities, social sciences, and management—people who are better prepared to meet the needs of business, industry, and our whole society.

Athena was in full swing by the time Malcolm arrived in 1987. He started out as a work-study student. He then worked providing tech support for Athena's large, decentralized network that connected more than a thousand computers. Professors, students, and administrators used them throughout the institute's research and development, teaching and administrative enterprise. On the one hand, Malcolm was a troubleshooter, in much the same way that William had been with IBM. But MIT promoted Malcolm to the role of consultant. That's when he began to work well beyond that technician role he first occupied.

I would answer the hardest questions that came up, from the faculty, the PhDs, the postdocs, and the student body in general. And at that time we had the most sophisticated computer network in the world. The Project Athena system, they hoped, would stimulate us students to develop the next generation of applications, which they, of course, would then go commercialize. When you hear somebody with a problem ten times, you're like, you know what? I can write a piece of software that makes that problem go away. When you're immersed in this system twenty-four-hours a day, you come up with these creative ideas, projects, applications. And that's what happened. We had hundreds of tricks and hacks, and applications that we built constantly. We were building programs that would later become core applications [for future computer systems]. Search engines of all different kinds. Configuration tools. Diagnostics. Just a whole list of things from the grab bag of tools you would want when you're making your system easier to use.

Knowledge. Practical experience. Malcolm secured a wealth of both. He had access to the project's collaborators—IBM and DEC scientists, software engineers, and product specialists. Malcolm graduated in 1991, but despite all his experience, he walked away from MIT thinking he didn't particularly have a leg up in the job market. No one seemed to know when the revolution was happening. No one knew what it would look like. No one recruited for a computer revolution that was . . . about . . . to happen . . . maybe . . . soon.

The aerospace industry—which had long been a go-to for MIT's graduating engineers—had taken a nosedive in the economy. Malcolm noticed that the banks, Wall Street, and the big management consulting firms grabbed up all his classmates—anyone coming out with an engineering degree at the time. Computers? Yeah, they were changing. Changing fast. But it would still be another four or five years before even the prognosticators would know which way the technological winds were going to blow.

Malcolm wanted to go to graduate school anyway. He dreamed of Stanford, but he took a short detour to Japan first.

* * *

David was back in LA, pursuing his success. Then, one day out of the blue, his Japan days caught up to him.

It's 1993. A friend from Japan, a brother—who was actually another charac-ter—grew up in Sweden, but was originally from Namibia. He reached out to me and said, "Hey, there's this guy who's going to be coming up to America who's going to be studying—he's coming to the Bay area, he's going to Stanford. He's going to get a master's of computer science. He went to MIT undergrad in com-puter science. A very cool young brother. You need to meet him.

I said, "Fine." That was Malcolm. Malcolm then came—he was ten years younger than me. He came to LA with a motorcycle, drove up to Stanford. You know, the character, he had been on an MIT kind of exchange program for about six months in Tokyo, which is when he met this guy Kenny. Kenny put us in touch, and so now here's this connection to the Bay Area and to Stanford, and he was at the computer science lab in 1993. So I'd come up and visit him and he'd come down to LA a lot more often, of course, because he wanted to meet girls and he didn't want to stay up around geeks all the time.

Malcolm may have wanted to check out the young ladies in LA, but he also knew David had just as much interest in the machines and tools he and the rest of his geek friends busied themselves building up around Palo Alto.

We became friends. David was always interested in what I was doing. I didn't think it was particularly special. It was just what I was interested in. But to him, I think he thought it was . . . there was something about it. Something that con-nected to the future, and where we were headed.

Kamal, Making His Mark

One might think a black man would be ecstatic, or at least comfortable, working in the 1980s in a coveted profession, and at a scientific agency whose very job was to reach the highest of heights. But Kamal felt like he was slowly losing his soul at NASA's Jet Propulsion Laboratory (JPL).

The California Institute of Technology (Cal-Tech) is to rockets and space flight what MIT is to computers and computer networking. Cal-Tech aligned with NASA's Jet Propulsion Lab in 1958, and by the 1980s, it was difficult to tell where the soldered lines between Cal-Tech, JPL, and NASA began and ended. From NASA's very beginnings, enigmatic, number-crunching, and analytically driven black women had made their mark on the agency. They successfully propelled the first Americans into space.[7] The 1960s-era civil rights movement made integrating NASA one of the principal, but below-the-radar, battles.

NASA was the highest profile and oldest federal agency devoted to applied math, science, and engineering. It became stalwart in pushing the federal government's affirmative action and equal opportunity mandates through the 1960s and beyond. Over the course of the 1980s, the number of black employees increased by two percent at the agency, from fourteen hundred to a little more than eighteen hundred. The rate of increase was steady across all occupational codes and grades.

Kamal Al-Mansour, a UCLA graduate with a degree in political science, had become one of these employees.

My first job was working with one of the NASA centers. I was an alien—pardon the pun. Here I am fresh out of law school, twenty-four years old, not knowing software, not knowing computers. But I could tell that there was something going on. There were connections being made. There were technologies being discussed. And I didn't see anyone like me participating.

What Kamal did know was copyright law. He had given himself a crash course while getting his law degree at the University of California's Hastings College of Law, in San Francisco.

It's funny. I had a close friend—I had a friend I grew up with, his mother was my piano teacher. He was the quintessential Urkel. He was Urkel and a Panther at the same time. He was so advanced in math, at the University of San Diego, professors used several of his formulas in their textbooks. He was the creator of laser tag. In my second year of law school, he had an issue with the misappropriation of his IP. So, it's my second year of Hastings. I assembled a team that specialized in federal litigation, that specialized in Japanese litigation. So I fight this issue that was taking place in Canada and Tokyo. Over the course of five to seven days, I immersed myself in everything tech, to understand the language, to understand concepts, protocols . . . everything I could. That's how I started.

Kamal went to work at JPL in 1983. It operated under the auspices of NASA but Cal-Tech managed it.

You had the affiliation with CalTech, with the Department of Energy. You have NASA sort of as the conduit, developing unmanned exploration. Everything was about software. I was negotiating technology transfers between Cal-Tech, with start-ups, a lot of start-ups in those days. A lot of the agreements were agreements created from scratch, because licensing was a brand-new concept, that software was intangible. You couldn't touch it and couldn't see it. It was really at that early stage of software licensing and technology transfers.

This was not unlike what was going on with Malcolm and Project Athena at MIT. When a university invests in developing a commercial product, the product itself is transferred to the commercial entity, if that entity wants, and has the capability to bring, the product to market. A legal contract governs the transfer. The agreement specifies things such as how much the commercial entity will pay for the product, who will own it, what rights the university and company retain, et cetera. Kamal negotiated these kinds of agreements on a daily basis, between Cal-Tech and its industry partners. In any contractual relationship like this, there is something to be gained by all parties involved. And, all parties share responsibility (whether they legally assume it or not) for the product's development— and, how people use it.

Keep in mind, Reagan was president at the time, and many of the projects I was working on at the time were in space. It was conflicting for me doing a gig that was supporting missiles in the sky, and on the other hand trying to find my own identity and culture in the world.

JPL was somewhat hostile in many ways and I would come home each day [thinking], What did I accomplish that benefited people like me? And every day the answer would be "Nothing." And in many ways, it was driven by a sense of wanting to accomplish something that benefited people like me. And just being somewhat insightful and attuned to what's going on, I could tell based on my experience doing these projects [at JPL], that there were very few of us doing anything in tech. And there was no market and nothing that existed. So I picked up a magazine. I'd see something and say, "I want to do that. I need to learn that."

I could write to the Organization of African Unity, and of course their stuff was dated. I could reach out to different universities and different ministries... try to find out what's going on out there. If I'm a business person or I'm a student seeking to go behind American universities and maybe branch out... I want to go to Africa? I want to go to the Caribbean? Where do I go? What do I do? Is there a single source of data that I can access? I was looking at my life, inward and outward.... Those were unique times.

But out of that climate came a calming, central focus, which was Afrocentricity.

It was a beacon. I was responding to that beacon, and I wanted to create a dig-ital bridge to what I thought was a beacon that many people were interested in.

* * *

The 1980s began to trail off. The 1990s were about to come roaring in like a lion. The Vanguard—they each led separate lives. Lee, David, and Kamal worked on the West Coast in Los Angeles. Malcolm was just a few hundred miles north in San Francisco. Ken Granderson and William clustered sepa-rately in the greater Boston area. Tyronne held fast down in Virginia, as did Derrick, down in Georgia. Others, like Ken Onwere in San Diego, Farai Chideya in Boston, and Anita Brown in Washington, DC, would later join the Vanguard's ranks. None of them (except David and Malcolm) even knew one another at the time. They had no way to know that their paths would soon cross. Physically, virtually, or both. None of them knew they stood on the precipice overlooking an open battlefield. Each of them prepared in their own way. But none of them knew exactly what they would be facing. Nevertheless, they were about to lead their people into a new technological theater.

Chapter Five

BLACK SOFTWARE COMES TO CAMBRIDGE

By the late 1980s, the black community in the United States had developed a sizable middle class. More students attended college. More graduates started careers at higher rungs on the professional ladder. Folks started businesses and built wealth.

But there was a trade-off. The strong community ties that had once propelled local and national civil rights actions, and spurred community uplift and provided social support—the ties that collectively vaunted community leaders to city council seats, mayoral offices, and the US Congress—began to weaken, fragment, or disappear altogether.

Kamal felt this isolation. He still hustled software contracts for The Man, out in the San Fernando Valley. He felt disconnected from his community and his roots, estranged from his brothers and sisters of color in LA, torn apart from his ancestral home, Africa. But Kamal made knowledge his business. He knew how to look for the answers he needed. When at first he did not find them he looked further. And deeper. What he didn't discover, he decided to create.

I started to build a database of any kind of information that could lead folks to invest in Africa and the Caribbean, understand the Caribbean, understand

Africa. That's what led me down that path, it was out of pure personal necessity. I had to imagine that there were others like me, embarking on the same journey or who needed to know that the journey was taking place.

Kamal toyed with computers. But when he started his quest, he retreated to his notebooks, not his keyboard or his word processing program. He carried these fat and thick notebooks like he was a prophet, and they were his holy books. He filled their pages with random thoughts and ideas, facts, and insights from the people he talked to—revelations gained from new relationships. Many of them sparked after he converted to Islam. He recorded news he watched, commentary he read, people he observed when he traveled. He collected stuff along the way: pictures, photographs, graphic designs, and those notebooks. These artifacts that he collected became Kamal's knowledge base.

Kamal was determined not to let the bastards around JPL grind him down. He looked for a way to lift himself—and, he hoped, his community—up. He started with what he knew. He knew his people. He knew the law, and the contracts that governed new business ventures. He was getting to know software. But he knew enough to know that African Americans *were late in their appreciation of science and technology*. He knew that just from looking around at the JPL ecosystem. Relatively few people who looked like him occupied or aspired to find computer and technology industry jobs.

On August 6, 1989, Kamal made his first move. With serial number 73805341, and registration number 1623958 both in hand, Kamal Amir Masiah Al Mansour filed *CPTime* with the US Patent and Trademark Office. He used his new trademark for the first time just a week later. On December 15 that year, he used it to sell his first new product. Kamal had packaged a clipart collection called *CPTime*. The art featured people like him: black people, Africans, African Americans, Muslims, people of the diaspora.

Colored people's time. We're late! I was thinking of it more as an affirmation that this is our time. This is people of color time. Time to move forward with technology in a positive direction.

Kamal's first step forward meant he no longer had to step foot inside JPL. He landed a new job, with GTE Government Systems Corporation. The company built telecommunications systems for the defense and aerospace industry. They located their headquarters in Needham Heights, Massachusetts, a Boston suburb. Kamal's wife, Angela, was a software developer. She had earned her degree in information systems from San Diego

State University. Both eyed new jobs and new opportunities. In 1988 they departed, together, to the other side of the country.

They could have chosen to live close to GTE, but Needham Heights's demographics left something to be desired. The chance they would ever find a black neighbor? About as good as electing a black president of the United States at the time. Which is to say, they had little chance in hell. Kamal and Angela chose to go where they would feel more comfortable. They decided to find a place where people would be more comfortable with them. William had made the same choice in 1980. Ken did the same after graduating MIT. Each of them made Roxbury home.

CPTime remained Kamal's hobby. He still had to work the job that paid the bills, and his frustration continued to fester. He continued to see software building new worlds all around him. But their architects, he noticed, made little effort to make a place for him, or his people, in their worlds. Then, the day came when he witnessed on television the very thing he knew, but did not want to fully accept.

I was watching a PBS show late one night and it featured a Princeton professor. I can't recall his name. And he was showcasing a software application called Culture. *And I kept watching and looking and I saw Michelangelo, references to Rembrandt....I kept waiting and anticipating that at some point it would make some reference to Africans, and there was none. And it convinced me that there was a professor going around pushing a software program about culture without any reference to the original culture. And it became my personal mission to develop a software program to counter that and to promote, if you will, or describe for the most part, African culture.*

So, the next day after I saw the program Culture, *I started a crash course in my townhouse in Roxbury. I upgraded to a Mac Plus.*

Among its many features, the Mac Plus included an astounding amount of random access memory (RAM). One megabyte, upgradable to four. This gave Kamal as much processing power and speed as one could find on any personal computer at the time. The Mac Plus also came with an SCSI port. The Mac Plus allowed users to attach external devices, feed input to the computer, receive outputs from the machine, and store them on external devices. This was something useful if one wanted to, say, transform analog content to digital, to burn content onto a CD-ROM.

I spent literally forty-eight hours straight trying to make something....Click a button, insert graphics, etc. etc. I took all the notebooks I'd been filling out with data. I started with a few pages. With Mac software, started with something that resembled software. I called it Mac Africa at the time. I kept building and building. A few months later, I had something that resembled software. The creative side was protected by copyright, trademarks. I had created packaging, which was crude at the time, but functional.

It was time for Kamal to really try to sell something. Before he had even left Los Angeles, schools in South Central's Compton School District had already purchased *CPTime* for teachers to use in the classroom. But now, Kamal had designed even more. And he was beyond eager to publish it to the world.

I thought maybe I should try visiting an AfAm book club. I was just selling it by word of mouth. I also wrote what I thought a press release might look like. So I sent out a press release to a local newspaper. Next thing you know I changed the name of the software package to African Insight. I had my first clip art—CPTime clip art on one.

* * *

Word spread. Kamal's zeal to broadcast his work even further serendipitously led Kamal to a computer store across the Charles River, owned by someone who looked like him. It was a business that perfectly communicated his new brand. He arrived at MetroServe Computer Corporation, a minority business enterprise and computer consulting, hardware, and software company located at 80 Prospect Street, Cambridge, Massachusetts. It was owned by William Murrell III.

William was always ready to shoot the breeze with any customer who came walking into the store. But he could not be more excited when Kamal showed up.

This dude came from Los Angeles. He walked right in the front door and said I've got something to show you. He said you got a store. I got a product. Kamal Mansour. He called his company AfroLink. And he had just completed a job with Mashari Medical College. They hired his company to create material, images about black people's lifestyles, to incorporate into an HIV-awareness program. So he had this whole library collection of like Afrocentric graphics. And then he converted that into a digital form and then made things out of it. So if you wanted to buy some black church scenes, you could . . . some black business scenes, you could. . . . And that was his project. That was his first project. I called it black software.

Back in Los Angeles, Lee had once reflected on one of the keys to his growing success. He had made *RadioScope*, his nationally syndicated radio program, successful by remembering that *content is king*. But he also realized that *distribution is key*. William understood this concept well.

At that time, you know, the only way to sell software, because everything was on disc, and we're talking floppy discs, we're not even talking CDs. They weren't really big yet. The only way to sell them was through a retailer. There was no other distribution. The idea was, you can get a reseller, stocking reseller, then you can put that down at these locations. And the whole idea was that you had walk-in traffic, you had an immediate customer. You just had to put it in front of them.

* * *

Kamal had put his software in front of a lot of people. And word got out quickly.

The phone rings one day, and it was Dr. Molefi Asante.

Now, to Kamal, picking up the phone and hearing Dr. Asante's voice on the other end was almost like talking to the prophet himself. Like Kamal, Asante graduated from UCLA (with his PhD). With others, he had pioneered the field of black studies. By the mid-1980s Asante was the chair of the African American Studies Department at Temple University in Philadelphia, Pennsylvania. It was one of the few departments where you could get a doctorate in the field. More significant for Kamal, Dr. Asante had founded the concept. He had written the first book and become the chief evangelist for that calming, central focus, that beacon that Kamal found when he struggled between the world that JPL built, and his own, still nebulous sense of identity. That beacon was Afrocentricity.

Dr. Asante asks me on the phone:

What's ready?
I have clipart, I have history.
What price? When is it ready? Keep up the good work.
<click>

It was that day that I said okay, I'm in the software business. That day I started AfroLink software as a business. That day I kept packaging software. Then I started manufacturing software in my townhouse. Then I started getting letters in the mail with checks. It was just going, going, going. It just took on a life of its own.

Time had come for Kamal to say goodbye to GTE. Goodbye to Needham, and Roxbury, and Cambridge. And to William, who would never forget the man who first walked through his door peddling black software. Or ethnic software. Or blackware. People started to call it different things.

* * *

Kamal began to build and distribute a new knowledge base by, about, and for black people all over the world. At the same time, a conversation had already started to percolate in the new parallel place called cyberspace. Was Usenet ready to provide a platform for users to talk openly and honestly about African Americans? That was the question. In March 1990, Eric Deering opened the floodgates.

From edeering@oracle.com

Flags: 000000000201

Newsgroups: news.groups,news.announce.newgroups,soc.culture.asean, soc.culture.china,soc.culture.latin-america,soc.culture.sri-lanka,soc.culture. hongkong,soc.religion.islam

From: edeering@oracle.com (Eric Deering)

Subject: CALL FOR DISCUSSION: soc.culture.african.american

Followup-To: news.groups

Summary: Creating a new newsgroup: soc.culture.african.american

Reply-To: uunet!oracle!edeering@ncar.ucar.edu (Eric Deering)

Organization: Oracle Corporation, Belmont, CA

References: <9001050517.aa24034@looking.on.ca>

Date: Tue, 13 Mar 90 01:42:07 GMT

Keywords: posted

This is a formal CALL FOR DISCUSSION about the creation of an unmoderated group to discuss issues and events of concern to African Americans, as well as people of African descent residing in the general vicinity of United States (i.e. U.S., Canada, the Carribean [sic], etc . . .). The proposed name of this group will be Soc.culture.african.american, and it will be a forum of discussion for topics usually attributed to the regular soc.culture groups (i.e. sports, music, politics, business, the media, history, the Arts, the sciences, etc...).

Because of the great diversity of African descendents [sic] in the Americas, it is difficult to create a newsgroup that is all things to all people. The primary focus of this group will be towards those descendents [sic] of the African diaspora living in the United States and countries adjoining the United States (i.e. Canada and the Carribean [sic] islands). It is hoped that someday other newsgroups of a similar nature will be spun off from this one.

So far, there has been a lot of interest in creating this group, via local discussion on soc.culture.african and various emailings.

Right now, the group is to be unmoderated unless someone would like to act as a moderator. Unfortunately, I cannot commit myself to the task, but if someone is willing, and people want a moderated group, please volunteer.

As per the guidelines, discussion will last at least 2 weeks, the discussion being held on news.groups, NOT soc.culture.african. Please post ALL future comments to news.groups and not to me. Discussion will begin today, Monday, March 12, and will end on Tuesday, March 27 at midnight. Afterwards, a call for votes will be made by me.

Thanks!

Deering, presumably of African descent himself, informed potential voters about the rules that governed whether to launch new newsgroups. Deering also laid out how the new forum should operate. That is, if the participants in this idealistic online democracy were ultimately to vote in favor.

But the discussion sparked tension from the very beginning, tension as thick as America's historic relationship with black people. It pervaded the conversation. No matter which direction it took, ultimately Deering would call the questions and put them to a vote—but not before the group's members had thoroughly perverted the proposed new group's original intentions. Perhaps he did not see it. But Deering neglected to highlight the discrepancy between the originally proposed charter, and the one he sent out for the vote.

Deering—and presumably those whose voices he channeled—wanted to start a forum to discuss "issues and events of concern to African Americans." Somewhere along the way, that got translated into "issues and events of interest to African Americans." The subtle change crushed a whole world of intentions, motivations, and possibilities. Nevertheless, the decision determined what this forum would become: 198 subscribers to the soc.culture newsgroup voted on this version of the charter, with 174 yeas and 24 opposed. Deering followed proper Usenet parlance. He published each voters' name, and how they voted. But the outcome was certain and clear. The people made history. The online world now had an open invitation to talk about African Americans. Some of the people doing the talking might even be African Americans themselves. The process was democratic, open, fair, and transparent—all the ideals people imagined this new medium would possess.

Yes, the people controlled their own fate. And the flame wars on soc.culture.african.american ignited on day one, despite the fact that the new group's charter tried to prohibit it. The participants scorched the earth, made it virtually uninhabitable for any black person to survive without their intelligence, morality, political interests, even their very identity being demeaned, called into question, or dismissed. Here these people were building a so-called new society online. They wanted to talk about issues of concern to black people. But almost inevitably they began to regurgitate the stereotypes that had dogged black people since they arrived in America.

This almost certainly went unnoticed by the vast majority of those already in, or headed to, cyberspace. But this reality awaited legions of black folk about to enter. Among many other things, this reality brought a new sense of the urgent need for the work that Kamal had begun.

* * *

AfroLink Software Inc.'s new cofounders had packed up and moved to Renton, Washington. Kamal worked on AfroLink full-time, from home. Angela spent her days in Bellevue. There she worked at Microsoft, the PC company that founders Bill Gates and Paul Allen had taken public in 1986.

The first thing that Kamal did for his newly launched business was to create a new distribution channel. He launched the *CPTime Online* BBS. But unlike many BBSs at the time, *CPTime Online* charged users a $39 subscription fee. By this point, Kamal had added a number of new black software products to his inventory. His *CPTime Clipart* package sold for $34. African/African American Insight—which featured information about Africa—sold for $49. And you could buy a new question-and-answer series about black history. Kamal called it *Who We Are*, and sold it for $69. Kamal distributed them all on CD-ROM.

By the early 1990s, when the Seattle-area newspapers and national black magazines like *Emerge* began to catch AfroLink's wind, only thirty subscribers had tapped into *CPTime Online*. But the CD-ROM by mail trade began to skyrocket.

* * *

About a thousand miles south, Malcolm commanded a bird's-eye view. His Palo Alto perch revealed the new and unfolding technology landscape.

When I first met David late in 1992, he didn't really know much about computers. We had a conversation about multimedia—primarily the CD-ROM—because that was really the best way to deliver a rich experience on a computer. You can pack a lot of data onto a CD, and of course present that on a screen. David took that seriously. He had all these clients, and they're in the content creation business. So he thought, maybe there's a way to turn that into a business using this new technology.

The moment had a deeper impact on David than it did on Malcolm. Malcolm had been looking at this kind of stuff for years. It was nothing special. It was just one of the many developing technologies strewn about his Silicon Valley world.

So as you'll see from the way I talk, I'm a very frank guy. So we're going to talk about this business the way it is. So me and Malcolm go back and forth, hang out. Of course I'd go up north, and one time in particular, I went to visit him, and we were seeing these things. He showed me these things called, at the time, CD-ROMs.

And a CD-ROM was basically just "you stick this in your computer and images would come up," data. It was the first time ever. It was kind of a cool thing that you could actually get more graphics out of your computer if you were looking at it, and see and hear sounds and stuff. So the way they were doing it first was an e-commerce—well, not they....In general, what I came to find was that most guys were trying to put—most companies up here in the Bay Area, the computer companies that were in this space, were trying to put catalogs, or the Yellow Pages on these CD-ROM disks.

Then, all of a sudden a band by the name of the Grateful Dead somehow got invited to do a CD-ROM and they were the first to do a CD-ROM. The sound and music and pictures and a little video. Then this guy by the name of Prince did it. So I was like, "Well, that's exciting!" So now I see an entertainment connection to what my friend up at Stanford was doing, my young friend. So I started, then. I joined the Beverly Hills Bar Association. I then started a little kind of group within it. First, the international law section, I revitalized that. But then I created this multimedia law, because that's what I had heard my friend up at Stanford call this thing.

So he was like, "Okay, cool." None of the white boys knew what that was. No one anywhere. But it sounded different and kind of cool, so I would make sure that all of my record clients had multimedia rights written into their agreements. And the record label is like, "Sure, whatever the hell that is." They had absolutely no plan. So all of a sudden, about a year later, it started to hit that this thing called CD-ROMs mattered, and more and more artists wanted it. So every artist was coming to their label asking, "Hey, part of what you're going to do is sign this record deal, but you're also going to launch a CD-ROM, right?" So they looked up and they saw several of their artists had already—they had given away their multimedia rights for several of their artists. So, they were like, "Who is this guy and how did he even know all this?"

The content on the CD-ROMs—obviously it was really clunky, so there would be music, but there would be maybe some kind of a short video or certainly pictures. Lots of images of the band performing, or the band in the background. They kind of would float around on your—you know how you may have—what do you call it? Your screensaver on your desktop. That was the level of the technology at the time. But having pictures of Prince sweating or crawling across the floor or sticking his tongue out, that's typically around the background while, let's say, Purple Rain *or something, right?*

David was right. Multimedia was the wave of the future. He reaped its benefits. But Malcolm knew it was time to have a talk with his friend.

David came by the campus one day, and I sat him down. These were the early days, and so all of the computer terminals for the graduate students at Stanford Computer Science Department were in the basement. We called it the dungeon. I brought him down there, sat him in front of a computer. I told him that I thought that multimedia was a dead business. The future is in connected computers. Not selling media printed on a CD or DVD.

Kamal would have begged to differ. He had just reached a milestone. He sold 2,500 copies of AfroLink software, netting him about $100,000 after just his second year in business. And that was just the beginning. On February 28, 1993, Anita M. Samuels of the *New York Times* wrote a column featuring AfroLink.[1] Anita had been eager herself to bring at least a hint of black life to the pages of the newspaper. She highlighted Kamal's *Imhotep* program, a collection of information and news about diseases that specially affected black communities. She also spotlighted his *Africa Insight*, which included a torrent of information about black issues. It informed folks about everything from black colleges to black elected officials. AfroLink's *Who We Are* commanded the greatest sales and Kamal had updated it with animation, and more questions and answers to fuel student lessons about black history, culture, politics, and more.

Things blew up a little more. And then schools, universities, just every single day I'm getting orders and it's just growing, growing, and growing. One of the things that I thought was interesting was that many inner-city schools in Cleveland, Cincinnati, Detroit were huge customers. What was funny, many of those customers in those days, it was their first time having computer labs in schools. It was their first time they were buying computers—and they said it was to use my software.

By this time AfroLink's software catalog included several volumes of clipart; history programs; educational games; and language training programs in Arabic, French, and Swahili. It also included the digital Koran and other electronic books. Kamal stacked each package with graphics and multimedia interactivity.

Kamal's multimedia success propelled AfroLink. But others had begun to see the same future that Malcolm had envisioned, one where computers facilitated human connection. And as 1993 began to fade, whispers began to fill the air that somewhere in that still-nascent computer network, there was not just a place for everyone. There was a place for people like us. Black people.

Figure 5.1. Kamal Al-Mansour, founder, Afrolink Software. Courtesy of Kamal Al-Mansour.

Figure 5.2. Afrolink Software promotional material. Courtesy of Kamal Al-Mansour.

Chapter Six

THE ELECTRONIC VILLAGE
NEEDS AN ORGANIZER

By 1993, the Internet had already produced many paths for black people to find one another, and many ways for us to find the information that matched our diverse interests. Bulletin board networks. Usenet. Bitnet. That doesn't mean they were easy to find, or that it was easy to find one's way from one to each of the others. The electronic village needed an organizer, an organizing principle.

* * *

AfroNet.
what is it?
(i'd like to know …)
Signed, B. Radney

Hi All,
Is anyone out there familiar with a computer network called Afronet???
Just asking!
Peace, Richard
If anybody does find out about this Afronet, drop me a line too...
—Peter Jebsen
I'd like any information anyone can give me about existing BBSes that cater to african americans, our concerns, our culture, our health, etc. These can be public

or private, including forums on existing commercial services such as CompuServe. Also, whether you have any information or not, let me know if you would like to see one, or be likely to use one. What would you like to see? Things I'd like to see are

Job postings

General Chat

Real national news (e.g., informed, local insight on what shows up in the media)

Music and movie reviews

Networking information

Business information (How to start one, opportunities, . . .)

I could keep going, and I haven't even gotten to the usual BBS stuff like files and pictures. Please let me know what you think, and respond to

   d . . . @fns.com

I'll post some indication of the response to this query in two or three weeks. Thanks.

David Jarrett

—Disclaimer—Whatever's above in no way represents the views of my company.

They'd be horrified.

At long last, someone bridged the BBS-Usenet divide with an answer.

KEN ONWERE

10/7/93

I finally made it to Usenet. After spending a good deal of time wandering around PC-based BBS message boards and echomail conferences, it sure is time for a change.

Let me introduce myself. I am Ken Onwere, an American-born Nigerian, and a graduate student (Radiological Health Physics) at San Diego State University. I am quite active in BBS-based networks such as fidonet. As a matter of fact, I am in the midst of a new project called Afronet. Afronet is essentially an e-mail backbone for distributing conferences with African-American themes. These are conferences, which due to message volume or participation level, do not qualify for transport over established e-mail backbones such as fidonet and usenet.

Afronet is an all-volunteer project, with over 20 nodes across the US and Canada. Of course, behind every successful project, are dedicated people. As such, Afronet needs volunteers who are willing to setup mail hubs on their PCs,

workstations, mainframes, you name it. If you can help, please send a note to
my e-mail address, or call (voice) at 1-619-594-3202.
BTW, has anyone considered creating a newsgroup which focuses on the use of
e-mail in the African-American community? If you have, let's talk.
Regards,
Ken Onwere
Dept. of Physics
SDSU, San Diego
California 91282
soc.culture.african.american >
Afronet newsgroup
1 post by 1 author

Ken was willing to waste one day, and nothing more. When Usenet echoed
nothing but his own voice, he steamrolled ahead.

Ken Onwere
10/8/93
[Article crossposted from news.groups]
[Author was Ken Onwere]
[Posted on 9 Oct 1993 03:55:24 GMT]
PS. This is not a request for discussion.
I would like to make contact with persons who wish to help create a news-
group for the Afronet project. Afronet is an nationwide e-mail backbone for
distributing conferences with African-American themes. At the moment, we
have a bbs network in place, across the US and Canada. However, we would
like to create a newsgroup under the comp.org hierarchy. This newsgroup will
serve as a source for news and information about the Afronet project, for
Usenet users.
Afronet is an all-volunteer project, we do welcome any assistance. We need per-
sons who have created newsgroups, or are familiar with the Usenet guidelines
for newsgroup creation. We are in no hurry to make mistakes, as such we antic-
ipate the process to take at least 9 months.
If you're willing to take up the challenge, please reply. Afronet is a small
project, as such we cannot guarantee that things will run smoothly. However,
we are a dedicated bunch, and we will work very hard to see this project
through.

Regards,
Ken Onwere
Internet: c...@radphys.sdsu.edu
Voice: 1-619-594-3202

When Ken Onwere set Afronet in motion, he was a student in the biomedi-
cal sciences at San Diego State University.

I was a teaching assistant. One day I figured out I could buy a PC, install a
modem, and dial into the college's mainframe. I could not only put in students'
grades but I could also run some of the lab's experiments on the mainframe, from
the comfort of my dorm room. And it was like the future opening up to me. This
technology was going to be great. I didn't know how it was going to turn out, but
it was like, whoa. This is something. Then before long you had the Compuserves,
and the AOLs came online. I could see something big was about to happen. Then
I remember one day I sent an email through Usenet to someone in Australia.
Typically, you send email it takes like a few days to get a reply, because your sys-
tem has to dial a system, which has to dial another system. . . . But in this case, I
sent this email and a reply came back in ten minutes. I said to myself, something
is changing.

Those changes led Ken to develop Afronet. Shortly after his epiphany, Ken
graduated with his master's degree. He experimented for a short time with
starting his own ISP, but realized it was not financially feasible for him. But
Afronet? That was doable. And to Ken Onwere, it would be very much worth it.

I felt that at that point in time, for me, it was a technical hobby. There were very
few African Americans working with online technologies. At the time, the Internet
was still just an academic research network. It was only visible in academic insti-
tutions that were linked to the D.O.D. network, ARPA. Afronet for me was, "Let's
get likeminded Africans and African Americans in the US, in Canada, who were
already active with online technologies, and let's see if we can come up with a net-
work," which was in essence a hub to exchange messages, emails, electronic bulle-
tin boards about topics of interest to us. For others, it was activism. Post–civil
rights activism. But for me, it was, "Let's get Africans and African Americans
interested in the technologies associated with people in my world," so to speak.

There were other people who already had an online presence, that catered to
African American culture, whether it was activism, social life, you name it. But I was
the first to step forward and say, "Let's register this name, and use it as a front, a hub,
to coalesce the different ideas that had been manifesting in online bulletin boards."

You had, for example, a lady named Idette Vaughan, from New York. You had Art McGee on the West Coast. They had their own electronic platform, either on the BBS or Usenet. So I just said, "Let's just interconnect our different ventures, and let's call it Afronet, under this technological umbrella and anyone can determine their level of participation or interconnection." Everyone still continued their own projects or ventures, with Afronet as a cover. My participation was short-lived. After a few years I turned over the reins to Idette Vaughan, who was on the East Coast.

When Derrick Brown walked into that Clemson cafeteria, he had looked for a place where he would feel comfortable. A place populated with people like himself. A place surrounded by people who shared at least some of his experience, if nothing more than just being black. It was a space that he and others like him came to not only cherish, but desperately require. Their social, emotional, and even their professional survival depended on it.

The early 1990s Internet had become a cafeteria, filled with the text-based musings and rantings produced by the generations of Usenet and BBS users from which the Internet evolved. It had become a place where you could find and sample expressions on just about everything you wanted. Porn. Politics. Suicide. Aliens. African Americans. Anarchists plotting world domination. And from day one, this environment was hostile to black people.

Number: 105/105 [yy Anarchy (N) yy]
Date : Sat 19 Oct 1991 2:13p
From : The Walkin Dude #470
To : Macgyver
Title : Re: CIS Hack
>> Reply to message 90
Hey Mac, I'd like to hear more about hackin CS... I'm all ears dude...

A thread like that in a chatroom might generate any number and type of responses. One of them came from an anarchist hacker who referred to himself or herself as *BeZeRKeR*.

Guide to Anarcy
    If you truly want to be a great anarcist there are a few guidelines you must live-by. These are 10 undeniable things that should be a part of your life if you truly want to fuck with society. There is always room to expand on the ideas and sick plans of Chaos presented in this little creation of my

obviousy demented head. Well here they are, don't doubt there potential to mold you into a sick, and troubled individual.

...10. If one day any of you succeed in a plot for world domination the follow steps !!must!! be implemented.

B>The seperation of all niggers...into work camps on the continent of Asia.

You might find your way to other such treatises that revealed the thoughts, feelings, and beliefs that early BBS users ported into the Internet.

Toxic Shocke Presents:

```
_____ /\ /\ _____
_____|)--<||> 50 Uses for the Household Pussy <||>--(|_____
) \/ \/ (
```

A 12/04/89 by Gross Genitalia A
Head-to-Hole Head-to-Hole
Production Toxic Shock File #3 Production
[Centre of Eternity.....40 megs.....3/12/2400 baud...........615-552-5747]

Pussy. Yeah, that nice wet soft female sex object us guys just fuckin flip out over. But what happens when that nice young female dies? Oh yeah if you're a fuckin gross-ass necropheliac you can keep her and use that stone-hard bitch as a fuck doll, but nah. You've loved her cunt for a long long time, why not keep it? Put that bitch to use. Here I present:

50 uses for the household pussy

>30. Wear it around your neck like black people wear the Africa symbols and Cadillac hood ornaments.

> 48. Go into Big Lots, where all the niggers shop, have your cunt handy. No not YOURS, I hope you guys don't have both! You know what I mean. Put the cunt on top of the manager's head then announce "Yo niggas, whassup? There's some free sweet pussy up here, come and get it!" See if the manager doesn't get mobbed with a thousand blacks trying to mob the son of a bitch.

BBS. Usenet. The Internet. Yes, they were creating a whole new world. But it wasn't a question about if and when racism would rear its ugly head in this new world. Racism, fueled by anti-blackness, was already there when it began. And if you were black, and online, your very emotional survival depended on your finding a respite in a new world that was, like the old one, built on, and permeated by white supremacy.

But Afronet became that virtual table where all the black kids could come to sit together. Afronet was where we could find our people.

* * *

As Ken Onwere explained, Afronet was a FidoNet system. SYSOPS controlled access to their respective boards. Most were free, and thrived on openness. If I knew your number, I was free to connect to your system and partake of all the network had to offer. But Afronet was a who-ya-know kind of system. When whispers and questions about finding and joining Afronet surfaced in Usenet, you needed its calling card. If you had it, then there was someone ready to answer.

John Alston
1/11/94
Hi David,
There is a network of BBSes called Afronet. I am a member and administrator. We carry a large variety of conference areas that are shared with about 26 systems nation-wide. You can call my system, The Vulcan-Net BBS and obtain a list of the BBSes, conferences and the possible location of a system in your local region or area!
The number is (908) 769-7882!
During the registration leave a comment that you read about it on the Internet and I'll set your access level accordingly!
Peace!
John Alston
1:107/918 Fidonet
jal...@cjbbs.com Internet

You could spot Afronet boards by their names. *Imhotep. Inglewood Rooftop. The Black Net. Nefertiti. Online in 'da 'hood.* Their names were SYSOPS's way of saying *this is a black neighborhood* in this electronic village. Afronet was a gate, keeping the race trolls at bay. For others, it was a door inviting black people to become part of a real community.

* * *

By 1993, Derrick had graduated from Clemson. He ventured a bit farther south soon afterward, bound for Georgia Tech in Atlanta. Derrick left

UNIVERSITY OF PENNSYLVANIA - AFRICAN STUDIES CENTER

AFRONET BBS List

This is a list of dialup Bulletin Board Systems/Services(BBS) in North America that are members of Afronet. This list has been gleamed from Arthur McGee's popular African-American BBS list (copyright Arthur McGee and Associates, 1993.)

```
This list is up to date as of JULY 29, 1994.
*
LEGEND: A=Afronet; B=BDPA; F=Fidonet; R=Rime
*
NAME                    NUMBER          BPS  NETWORK ADDRESS
*
*
AfraLink                (***)NOW-DOWN   ***  101:19/2(A)
Kinky Kumputers         (***)UP!-SOON   ***  101:13/136(A)
Black Business Net[1]   (201)836-0602   144  1:2604/315(F), 101:13/132(A)
Black Business Net[2]   (201)836-1844   144  1:2604/315(F), 101:13/132(A)
Boardroom(IABPFF)       (201)923-3967   96   1:107/819(F), 101:13/111(A)
Tracks From The Past!   (205)237-6074   144  101:13/134(A)
Invention Factory[1]    (212)274-8110   144  101:13/125(A)
Invention Factory[2]    (212)274-8111   144  101:13/126(A)
Invention Factory[3]    (212)274-8112   144  101:13/127(A)
Informed Sources        (212)281-9478   144  1:278/303(F), 101:13/108(A)
Poetry In Motion        (212)666-6927   144  101:13/130(A), #1064(R)
Humanity                (213)936-6009   96   101:10/5(A)
Alex's Place            (213)937-8734   24   1:102/744(F), 101:10/4(A)
Minority Affairs(BDPA)  (214)517-7254   144  1:124/3121(F), 101:19/1(A), (B)
Vulcan-Net III(IABPFF)  (215)223-2995   144  101:13/123(A)
African American[2]     (215)842-1560   144  1:273/957(F), 101:13/121(A)
African American[3]     (215)842-1561   144  1:273/957(F), 101:13/122(A)
Jailhouse               (305)944-6271   144  101:13/135(A)
Inglewood RoofTop!      (310)677-4007   144  1:102/355(F), 101:10/6(A)
Nefertiti(BDPA)         (312)488-8969   96   101:11/2(A), (B)
Wit-Tech(BDPA)          (410)256-0170   144  1:261/1082(F), 101:13/103(A)
Heritage                (410)732-3368   96   101:13/114(A)
U-People                (419)589-2310   144  1:234/62(F), 101:11/1(A)
Online In d'hood(IABPFF)(609)645-7080   144  1:2623/65(F), 101:13/118(A)
AfroConnections         (613)237-9531   24   1:163/511(F), 101:12/1(A)
Spirit DataTree         (617)983-9590   144  101:13/119(A)
Timbuktu                (714)638-2954   144  1:103/197(F), 101:10/3(A)
Imhotep                 (718)297-4829   144  101:13/120(A)
Treasure Chest          (718)525-5610   144  101:13/128(A)
Holman's World          (718)529-8890   144  1:278/511(F), 101:13/107(A), (B)
Utmost Quality Bulldog  (718)562-2745   144  101:13/131(A)
Ashanti Connection      (718)634-4175   144  1:2603/214(F), 101:13/110(A), 1718000(V)
The BlackNet            (718)692-0943   144  1:278/618(F), 101:13/104(A)
Systematic[1]           (718)716-6198   144  1:2603/505(F), 101:13/106(A), #5210(R)
Systematic[2]           (718)716-6341   144  1:2603/505(F), 101:13/106(A), #5210(R)
Spaced Out              (718)778-5637   24   1:2603/108(F), 101:13/116(A), 7:718/1(SACnet)
Cypress Hills           (718)827-4933   144  101:13/115(A)
Fort Knox               (718)969-6910   144  101:13/117(A), 7:718/4(SACnet)
Vulcan-Net(IABPFF)      (908)769-7882   144  1:107/918(F), 101:13/105(A)
Islam On-Line[1]        (912)929-1073   144  1:3611/21(F), 101:19/3(A)
Islam On-Line[2]        (912)929-2873   144  1:3611/21(F), 101:19/3(A)
Outside The Mainstream  (914)381-2390   144  1:272/74(F), 101:13/124(A), 1914000(V)
```

Figure 6.1. Listing of Afronet Bulletin Board System names, dial-up phone numbers, and other board characteristics, 1994.

Clemson confident in himself, confident in his craft, confident about his desire to serve others. This was the stuff people couldn't see just by looking at Derrick when he arrived on campus. What would he do in this sea of top-of-their-game engineers? How would he distinguish himself among its talented tenth? Black students were a minority at Tech, for sure. But there were enough of them around to be noticed. How would Derrick serve when his people had no need for a tutor? How would he help maximize opportunities for those who had landed at Tech precisely because they had the intellect, the acumen, the drive to pursue their own dreams—on their own?

There we were, a bunch of us from all over the country. Different fields of engineering and science. And when we got there we kind of all realized that all of us had gone to schools where we were pretty much isolated. Being students of color, in engineering and science fields. Now we were at Georgia Tech. We were all isolated—together. At first, the electrical engineers were hanging out with electrical engineers. Chemical engineers hang out with chemical engineers across the street. But we were all engineers, and we all interact socially, with the Black Graduate Student Association.

My first year at Tech I created a committee called Information Resources and Technology. My past was computer focused, but very database focused. I wasn't really into coding very much. In one of my classes at Tech, when we're supposed to be, we were going to become hardcore coders in that class. So I told my teacher I'll do this code stuff but I don't really write code. I'm more application focused. And at the time he told me, "If you're going to be in computers, man, you need to write code." I said, well, you're right but if I'm going to be in computers I think I'm not going to be somewhere on that lone level of building blocks and software. I don't like that.

So I created this committee on tech with the BGSA, and I said, okay, the focus of this committee, I'm going to get coders together with database people and email people. We're all going to show each other how to use this stuff, because it's going to be valuable in our research.

At the beginning, not every one of his BGSA colleagues picked up what Derrick was laying down. They didn't see how some tech committee's work was going to help them fulfill their academic goals. But let Derrick throw a party? That would help everything sink in. You see, Derrick was an engineer. He was a rising star in the BGSA. But Derrick was also a DJ, and his apartment was the party spot. And the only way to get directions to the party on the fly was to check your email! Print it out, and take it with you.

Derrick's parties provided a well-deserved respite. His BGSA tech committee had worked hard that year. Derrick's work—and no doubt his party skills—made him popular enough that the BGSA elected him president. As part of his transition from chairing the tech committee, Derrick summarized the year's accomplishments in an email sent to the following year's co-chairs.

WRAPUP.TXT
20 JUN 94
To: Alou Macalou and Carlton Riddick, Info Resource Tech Comm Co-Chairs
From: Derrick Brown, BGSA president
RE: Wrapup of 1993–94 Info Exchange Comm activities
Although I have already read your documented plans for next year, I thought that I would take some time to officially wrap up my activity as Info Exchange committee chairperson by making some suggestions for next year's activity and transferring a few files to you that should help with next year's projects.

The following are what I deem to be important issues the committee should plan to resolve next year:

- Increasing the number of BGSA members who access email and our newsgroups. The classes at the beginning of the quarter should increase usage a lot, but plan on stuffing boxes once or twice to coax people into activating their accounts.
- Devise a procedure to efficiently generate the BGSA mailing list. Perhaps procmail is the answer here, but we need as easy a method as possible. I'll be glad to provide the details of how I generated the list last year.
- Assist with the solicitation of a high-performance PC to use as a file server by providing details of how the committee will make use of these substantial computing resources that we will request.
- Continue to pursue using procmail as our mailing list manager; Adam Arrowood of OIT has offered his assistance in helping us come up to speed with the tool.
- Find someone else to maintain our mailing lists & aliases besides Russell "I crashed the hard disk & never back stuff up" Earnest, of Student Services.

I have included with this documentation a disk containing files that correspond to the following projects that were planned and/or executed by the Info Exchange committee during 1993–94:

- Creation of mailing lists (BGSA, sistas, Focus '94 attendees)
- Directory of Focus '94 attendees
- Establishment of two newsgroups, git.club.bgsa and git.talk.african-american
- Research for classes on using the Internet and Unix-based electronic mail
- Research for an efficient mailing list manager (procmail)

There are readme files located strategically throughout the previous directories on the disk. I also have hard copies of some documentation that should prove helpful in assisting with some of the tentative projects scheduled for next year. Unless indicated otherwise, these documents all appear on the disk. These documents include:

- 1993–94 Black graduate student directory
- Outline for the Internet classes that were researched in 1993–94

- Surfing the Internet
- Procmail descriptions and help files
- A paper I got from Russell Earnest on building listservs

Also, I direct you to the following helpful publications available at the 141 Rich:

- Zen & the Art of the Internet, by Brendan Kehoe
- Internet Gopher's Guide, by Paul Linder

That's a quick wrapup on what has been done & what needs to be done. As I said earlier, I have already checked out your plans for next year & they look solid. Try to make sure all the things that I have suggested here fit somewhere into your
structure.
DBrown

No one knew what was coming when Derrick first arrived at Georgia Tech. But if no one saw it by then, this memo made it crystal clear that Derrick was on a mission.

I gave that committee a couple of dollars and I said you guys keep building stuff. Let me know what you're doing and go ahead and tell me about it, and we'll see what we can do to get paid for it.

But Derrick had done more than drop a few bucks and send the committee on its way. Before he set them on their path, he packed as much of the BGSU's engineering power and diversity onto that committee as Tech had to offer.

Carlton, he went to a historically black college in Charlotte, North Carolina, and he was from New York. I think he was from Brooklyn. Alou Macalou, he went to MIT and he was from Mali in Africa. Paul Campbell went to Clemson with me as an undergrad. As a matter of fact, we knew each other in high school. He was my roommate at band camp one year in high school. But he's from Myrtle Beach, South Carolina, which is about five hours from where I'm from in South Carolina.

Larry Ward went to MIT and he was from the Caribbean, I'm not sure which nation, but it's Caribbean. Matthew Parker, he went to MIT, he's from Savannah. He went to MIT with Alou and Larry, so he always knew what was going on. He's the guy that taught us the meaning of the word diaspora. *I don't know if you have any friends like this, but he's like a preacher type of person. Very evangelical kind of guy.*

Leslie Gant, she went to Southern University, that's a historically black college in Baton Rouge, Louisiana. Camille Dozier went to Purdue and she grew up in

Indiana. She was the only black person I had ever met who grew up in Indiana. Derrick Coleman went to University of Kentucky. Marquis Jones was from Arkansas, went to Texas A&M. And Paul McLaughlin was from Case Western Reserve University in Cleveland.

All of us were so different. Whenever we came together and combined our mutually exclusive interests. That is what always provided the direction for all of our efforts, all of our conversation. I don't think we ever sat in a room and said, "Let's unify the diaspora." It was more like, "Here's what I'm interested in," and then another person chimes in what they're interested in, and we used to have these gender-based meetings. Like all the guys, we'd get together with the guys, and girls would get together with the girls.

We would all chart this course for the empowerment of all. And I don't know that we'd ever do anything with what we talked about, but the driving force behind all of the activities was just everyone's mutually exclusive interests, all combined, to do something that none of us could have done individually.

Derrick was modest, even as he embarked on a strategy that could lead to nothing else but something big. Big for Tech. Big for black people. Big for the diaspora. Big for the Internet.

* * *

Derrick and his information committee never slowed down. By the time everyone returned to campus in the fall of 1994, the committee was ready to share and showcase what they were learning.

We started teaching people how to access what we had been told was this thing called the Internet, and a text browser call Linux. It can do everything on command. You can execute FTP. You can execute Gopher. You can execute Binary Information Service. All the Internet protocols, you can do them on Linux. You can even start pulling up these HTTP sites that they call websites. Can't see them, it's all text. You go to Yahoo, it's all text and you had to be skillful in Unix, being able to write in proper commands, to get information downloaded, printed, and just have access to all of this stuff.

To the BGSA, we were rock stars. But we were also space aliens. So at every meeting we had a committee structure, we had like six committees. Each committee would give a two-minute report so that everyone could learn about what they were doing. So the Information Committee would stand up and talk about a website. We would just tell them about these things, real style, tell them like we're West African storytellers, telling them stories they cannot yet see.

And they believe us. But they're like, "We don't quite grasp what y'all are talk-ing about." You see, everybody's different at Tech. So we on that committee—we were more different than the different folks. So they're being told stories about a world that they aren't privy to, just try to grasp, you know, what's coming. We would take them to a computer lab, and they could see that computer. You type in these long addresses and eventually these pictures come back. And it's really slow, but it's working just enough that everybody gets really excited now. Because the people that you've been telling these stories to—they think we're crazy. But now we can show them, this is what Yahoo looks like. Look at what happens when you click here, and look at what happens when you click here. It's hierarchical, so you can keep drilling down. You can pretty much find whatever's out there.

It's at this point that Alou and Larry and the guys on the committee are telling me—we are trying to pull together, we're going out looking for stuff that's African. Because we were being ironic. If we could unify the diaspora, as it were, right here in Atlanta at Georgia Tech, where we're like the diaspora within the diaspora—that's just a beautiful metaphor.

So I'm like, "Yeah, it is." So Lou writes all the code, Larry is what I would call the editor-in-chief. He's writing brief descriptions of the site. He knows HTML, so he's building HTML pages. It's just like Yahoo, but instead of concentrating on the whole Internet, we're looking for the African-flavored content that exists, and we're using Yahoo to find most of that stuff.

And then we started to find a few things, and you start emailing the person who maintains that site. Then you start finding out more, and adding a few links. Two or three months, and then kind of add five or six more unique links.

Soon, the committee had a real vision. They had produced something tangible to show for all their work. Something that was legible to everyone, not just the computer geeks on the Information Committee. On October 23, 1994, the Georgia Tech BGSA announced that they had their very own page on the World Wide Web. They also announced their plan to build the Universal Black Pages. As the president of the BGSA, Derrick articulated the Universal Black Pages' early vision.

The main purpose of the UBP is to have a complete and comprehensive listing of African diaspora–related Web pages at a central site. Currently, the World Wide Web consists of many islands: each page addressing the particular interests of the sponsoring institution (or individual). A particular page may have links to other pages that are related; however, those links are far from being exhaustive or organized.

WWW: Georgia Tech Black Graduate Students Association

Figure 6.2.
Announcement of the creation of the Universal Black Pages by Georgia Tech's Black Graduate Student Association.

The Universal Black Pages (UBP) became the first black-oriented Internet search directory. By typing in a word, people could find information on cultural events, businesses, organizations, and more. Black users created websites with black-oriented content at lightning pace. UBP became both a map and a transportation vehicle. Collecting links to all of these disparate sites created pathways for other users to navigate and reach them. The ability to link came courtesy of Tim Berners Lee. He had used hypertext as the foundation for the Web that he created and gifted to the world. Using these links, black users searching for content for, by, and about themselves could not only see a pathway to reach each site but be transported right to them. Anywhere you wanted to go in the black diaspora. That was UBP's purpose. Unify all these disparate black sites and make them easier to find. It was to be as it stated on its page—*a guide to nearly everything networked and African, Caribbean, African-(North/South)American* online.

* * *

Thanks to the Vanguard—those I have named and those not revealed—black people stormed the World Wide Web shortly after it launched. We built websites, new outposts for a black diaspora hungry for knowledge about our origins and ancestry. We craved evidence that our prolific cultural productions were both real and valuable. We yearned for connection and communal support for our social, professional, and economic strivings.

Meanwhile, business boomed for Kamal and Afrolink. His black software company was poised to pull in nearly a million dollars in sales in 1994.

And like an island reprieve, rather than a ghetto, Afronet expanded its territory in the new electronic village. In less than two years, Afronet had organized itself into a network that connected black users from coast to coast, and from the South into Canada.

Ken had left the network by 1994. Afronet was growing, seemingly by the day. Art McGee, himself a student at the time, assumed responsibility for documenting the organization's growth and structure. He sent regular messages through soc.culture.african.american and other Usenet groups. He posted a permanent copy of the same information on his board, and later a website. He posted SYSOP names, board names, names of the zone coordinators from the West, Midwest, South, and East. He documented their boards' phone numbers, locations, as well as information about whether they were connected to Afronet, FidoNet, SACnet (Secure Automatic Communication Network), Black Data Processors Association (BDPANet), or other networks. McGee's post specified the specific BBS software the board ran on, whether Tri-BBS, RBBS, Renegade, Wildcat, or a multitude of others. If nothing else, Art was thorough, detailed, and serious about the work he did to document Afronet's growing footprint.

Afronet had its pioneers, Ken and Idette, but it also had stalwarts like Art, and similar BBS evangelists. But several other black professional BBS networks helped to power Afronet, including one created by John Alston. He was among the legion of black firefighters across the United States and abroad. They connected with one another and the Afronet through the International Association of Black Fire Fighters (IABFF). The association had built its own BBS network, which included several SYSOPS that were part of Afronet.

Earl Pace Jr. started the Black Data Processors Association (BDPA) for black IT workers in 1975 and expanded it into a national organization in 1978. By the early 1990s, BDPA operated another expansive network teeming with black technology professionals like Mike Holman. Holman had grown up in Queens, New York, and had joined that stream of MIT to Wall

Street graduates that Malcolm—now out in Silicon Valley—had once mentioned.

* * *

By late 1993, the calls began to ring out from multiple locations in black cyberspace, calling on us to continue to proliferate and reproduce ourselves. Others called to bring us together. Georgia Tech's Universal Black Pages started reaching out to different website owners. Its developers wanted to create a direct pathway between all of them.

At about the same time, Timothy L. Jenkins became publisher of *American Visions Magazine*. The Association of African American Museums and Smithsonian Institution jointly sponsored the publication. In it, Jenkins posted a clarion call to any and all blacks in cyberspace willing to read and heed its words.

Chapter Seven

WANT AD FOR A REVOLUTION

In several of my past letters I have promised the introduction of new technologies to facilitate dialogue with our readers beyond the printed page. American Visions is now on the threshold of implementing such a design, through a computer network called the "African-American Cultural Forum." This user-friendly system will allow the interactive flow of information between American Visions and subscribers on a wide range of interesting subjects.

The only requirements are a personal computer (PC) and a telephone connector (modem). With these two devices, a subscriber will simply call up the Cultural Forum's index on his or her PC screen and select the particular subject matter, reference material, article or publication for review. Subscribers also may wish to circulate materials, questions or comments through the Forum to generate feedback from other subscribers.

Think about it: An electronic inquiry from a schoolteacher in Los Angeles can be answered by a scholar in Atlanta and commented upon by a similarly interested amateur in Detroit almost instantly through such a network. Suppose someone wants to know the names of black composers of classical music during the 1800s or the text of the 1989 American Visions article on kente cloth. The Forum would provide the answer electronically.

The revolutionary potential of such intellectual synergy is that, for the first time, the African-American community will be able to enjoy a constant channel of information relevant to their interests from any part of the United States, as well as much of the world, without leaving the home or office. This form of electronic mail has the potential of bringing other subscribers and libraries of the world to our fingertips, as well as conveying original thoughts, "want ads" and questions to researchers and information centers.

The design of the Forum will require our collective thought and planning to realize the exciting range of the system's potential. Accordingly, we wish to convene the first national African-American Cultural Computer Symposium to explain and explore the available options for exploiting this new technology. We want to explain what we are planning, and we want to learn what information subscribers wish to access as well as publish through the Forum. Also we need to consider how all age levels, particularly youngsters, can learn to enjoy and benefit from such a technological breakthrough, with minimum costs and maximum satisfaction.

If you would like to attend the African-American Cultural Computer Symposium, to be held in January 1994 in Washington, DC, please fax me your name and address, along with a brief description of your interests and the uses you wish to have served by the Forum. American Visions' fax number is (202) 462-3997. The symposium will serve as a cultural brain trust, enabling us to collectively think through how we can ensure the long-term value of this revolutionary breakthrough in technology.

You need not be a computer whiz to contribute to and enjoy the session. Our interest is to broadly address the information needs of our community for the 21st century. The urgency for such a seminar rests on our recognition that the future in this field is now. This is the cultural revolution that American Visions is determined to influence and help build.

Mr. Jenkins had been one of the founding members of the Southern Christian Leadership Conference, a black civil rights organization formed in the late 1950s. Jenkins used his civil rights credentials, connections to the Washington, DC, black political establishment, and his entrepreneurial acumen to seize his position as publisher of *American Visions Magazine*, which began publishing sometime between 1984 and 1986.[1]

Jenkins's November call to the Black Internet was no accident. His vision was not a naive shot in the dark. He was a cultural visionary and architect, and November's pronouncement laid the foundation for a well-designed plan. Thus Jenkins began immediately gathering the brick and mortar he needed to begin building.

* * *

Black America heard Jenkins's call far and wide. Many answered—with some familiar names, including Ken Granderson and William Murrell among them. Ken had graduated MIT with a degree in electrical engineering but ended up going to work for a software company.

I'm working at a company called Phoenix Technologies, which was the first company to legally clone the chips that make a computer IBM compatible, and I was doing software testing. I got interested in programming, but since I was in the QA department, they weren't going to send me to any classes, so I taught myself how to program with some books and stuff and put out some shareware programs. These were the days when you could actually make money selling software to the public because people did not believe that software could be free.

So I did a couple of little shareware programs. There was a book called Windows Gizmos. It was a collection of shareware utilities edited by one of the writers from PC World, Brian Livingston, it was in computer stores, whatever. A little desktop utility that I made went in that book. Because of that and some other things—I remember it was in a book in France—I ended up with two thousand registered users of this $20 desktop utility, which earned me forty thousand bucks. Which is about what I was getting from my day job. So I quit my job and started programming full-time.

Ken started a business of his own in 1993 called Inner-City Software. That's how Ken Granderson found his way to Murrell's store, which had been steadily booming since 1983.

You know, he was actually a black man with a computer store, which was probably, you know, the only real computer store. When I say "real computer store"— I mean I've seen a couple of computer stores that sell more game machines, you know, and just do computer repair, which is cool. But William's store MetroServe— it was a competitive, it wasn't like a small—when I say quote mom-and-pop, I'm thinking in terms of a tiny corner store outfit where folks are just like, yeah, let's sell some sodas and beers. He had a very good, a very competitive, mainstream-looking computer shop. Like places that used to be around like egghead software. They were software really. But you could go into—I'm not good with square footage, but maybe a couple of thousand square feet, where you could look at machines and stuff.

Like Kamal back in 1989, Ken was a man whom William would always remember from the time he first met him.

I don't think it was random. I think it probably, well, well, he was an engineer. A software engineer. He was a graduate from MIT. Ken was the first person I ever saw program anything.

Ken and William both traveled down to DC for the meeting that Jenkins called. As did Mike Holman, who had been online since 1981. He had encountered his first computer when he was a student at Andrew Jackson High School in Queens, New York.

My first computer was, I can't even remember the name. I think it was by Toshiba. But it had like a 3.5" disk, and the hard drive had to be like, maybe 20 megabytes. The RAM, maybe 1 gig. Then, after that, then I got a TRS-80 from Tandy to hook to my TV.... The Toshiba had a built-in modem, 300 baud. That was the max, 300 baud. And then the TRS-80, it was external modem, I think 300 baud as well.

Mike discovered BBSs sometime around 1990 or 1991. Most folks stumbled onto Afronet roaming their way online through various networks. Unlike them, Mike ended up there through his professional association with BDPA, the organization that published print newsletters. One issue ran an article by Idette Vaughan, where she expounded on her wisdom of BBSs and Afronet.

Idette had been a legal secretary before she decided to start her BBS in 1989 from her home in Brooklyn, New York. She called it Blacknet, and she built it explicitly to make money. But making money for Idette was about making connections to, and brokering information about, members of the black community.

At the time, a lot of people said not many black people have computers. I just said, "We'll have to see about that."

By the time Mike Holman had found her, *Black Enterprise* magazine had reported that Idette had about six hundred monthly Blacknet subscribers, at $10 a pop. The subscription gave you access to black video games and screen savers as well as other content available for purchase. More important, the magazine explained Idette's business model.

Vaughan's Blacknet is linked to Fidonet, which is also linked to Afronet, which is a group of nation-wide, Black-owned BBSs. So, theoretically, if you list your BBS on the mail board of Blacknet, users of other BBSs such as Afronet, will have access to your posting as well.[2]

Michael followed Idette's business logic. And her community focus.

She actually wrote something on bringing unity within the BBS world. And that made me more interested in that and actually dial into her system as one of the first systems that I dialed into.

By 1993, Mike had started his own BBS. He named it Holman's World, and joined Afronet.

What I wanted to do was cover as many areas as possible, so I had a board for technology, a board for maybe sports—although I'm not heavy into sports, I wanted a board for that, so I created a board for that within the BBS. I had a file area because people are going to want to share files. So I had a file—people could upload files, download files. Creating that. Creating a user interface, in terms of what users I wanted. A lot of it had to do with design, in terms of what I wanted the users to feel when they get on the board. So a lot of it had to do with design work.

I wanted to do a lot of things. I was also thinking, in terms of my business, I wanted to be able to start, to find out about accepting payments to pay for things. So there was an application that allowed me to not only have a board, but be able to let people buy things from the board. So I actually—I had that in the BBS, where you could say, "Oh I like this file." And I say, "OK, it's only $10." And they'd purchase it. I'd then send the purchasers to another site, and I'd get payments however I designate—my credit card, I want to get payments on my checking account, whatever. And I would get payments from that.

Art McGee, Idette Vaughan, and a host of other Afronetters ventured down to, and out for, the DC meeting. Even Kamal flew all the way out for the event. It was the kind of gathering he had been hoping would take place for a long time.

* * *

Timothy Jenkins described what had taken place at the January meeting in his "Publisher's Note" in the May 1994 issue of *American Visions*. He called it an *Interactive Niagara Movement*, referencing the movement that W. E. B. Du Bois had started back in 1905 to fight segregation.

When the president [Bill Clinton] *emphasized in his State of the Union address that the "information super highway" was paved with good intentions, he neglected to note the potential detours and dead-ends that could lead the have-nots of today to become the know-nots of tomorrow. To forestall such a result,* American Visions *called for a national African-American Computer Culture Symposium. Those readers who accepted our invitation and that of the Congressional Black Caucus Foundation to brainstorm on how we should participate in the multimedia computer age included university presidents and administrators, executives responsible for major African American collections and archives, religious leaders, entrepreneurs, technicians, activists, elected officials, librarians, systems operators, schoolteachers, independent film producers and foundation executives.*

We braved ice, snow and the lowest temperatures recorded for Washington, D.C., in 100 years to attend the January 1994 Symposium. And what an electrifying session it proved to be, resulting in a multifaceted commitment to participate in the coming multimedia computer age by any means necessary. Keeping a steady pace from morning to night, participants reviewed emergent technologies, corporate players and government policies, as well as the demographic reality of our marginal cybernetic presence to date. All of the Symposium's speakers shared one trait: a depth of expertise and a burning commitment to change the way information science can be linked to problems' solutions. Written background materials included the detailed policy concerns of AV and CBCP, as well as descriptions of a broad range of interactive innovations for business, political organizing, distance teaching, entertainment, networking, travel and shopping. The revelations of the Symposium were that just because something is complex, it need not be confusing, and that enlightened self-interest requires that we choose proactive planning over after-the-fact complaints.

It was agreed to establish Symposium working committees to plan three days of workshops and industry demonstrations for a major national expo on computer culture. The following subject areas were proposed for sessions at the expo: (1) new technologies, (2) education, (3) labor market effects and opportunities, (4) business options and necessities, (5) cultural institutions, (6) entertainment and sports, (7) networks and bulletin boards, and (8) telecommunications governance, regulations and democracy.

I welcome additional volunteers to identify speakers, panelists, demonstrations and handouts, as well as proposed goals and timetables for community action.

We who were the victims of the technology revolution in agriculture and who were ignored by the Industrial Revolution cannot afford to be bypassed by the multimedia communications revolution inherent in the emerging information super highway. We must mobilize not just our personal energies, but those of any and all organizations and institutions in which we participate, to avoid our continuing characterization, lamented by W.E.B. Du Bois, as an "afterthought of modernity."

We must seize the time between now and the twenty-first century to prepare our children and ourselves not only to survive, but to prevail in the information age, just as the Niagara Movement resolved to ensure the fullest social participation early in the twentieth century. The issues of access and inclusion are the same, only now they're dressed in new technological clothing.

Late in May 1994, Jenkins traveled to Egypt. The trip provided him an opportunity to reconnect with his ancestral home. An opportunity to help his black *American Visions* magazine subscribers feel in touch with the land to which they were connected, but could not experience first-hand. There in Egypt, Jenkins sat, looking out the window. His eyes transfixed on Cairo, and the Nile River that ran alongside the bustling Egyptian city. Nostalgia set in, fomented not by memories of his own youth, but rather on the thousands-year-old history that lay bare in the Egyptian landscape. Its tombs. Its monuments. Its art. Its writings. He mused about the clear, but little-recognized history that connected that place, those people, and its artifacts—to the black men and women among whom he lived every day.

Pacing through Egyptian structures, streets and thoroughfares with cornerstones from 300, 950 and 1400, both BC and AD, provides a heritage of time and place that allows me to walk more heavily upon the earth and more confidently into the future, without the need for extravagant propaganda. I cannot help but believe that were the next generation taken by the hand through the tombs and temples, villages and urban centers, art museums and monuments, churches and mosques of the Old World, the experience would fundamentally alter its values.

Taken seriously, such an international exposure could compellingly redirect a generation's attention from the virtual unrealities of sports mania, fashion chic, gun dependency, cultural myopia and technology denial to the wide discoveries of reality-based imagination and cultural literacy.

In addition to his magazine, Jenkins had an eye on an existing powerful network—one that had already been built and existed online. That network had an infrastructure, powerful voices, expertise, and everyday people connecting every day. Afronet saw itself as a community first. But Jenkins thought he could leverage it for more.

* * *

Jenkins planned to recruit as many Afronetters as possible to join what *American Visions* planned to build. American Visions' leadership had chosen William as their emissary. But they had not revealed that publicly yet. Not even William knew. To William, *American Visions* was just a magazine. He'd been thinking about what he read in it. But he had no idea that the magazine's publisher even knew who he was. Until…

I was an Afronetter, had my own computer server campaigns. I was pretty busy. Very happy. And I was also an American Visions *subscriber. I read it. My family read it. It was a great magazine. So I made a phone call to* American Visions. *To Jenkins.*

William was forthcoming with Jenkins. He knew he had something to offer *American Visions.* Jenkins revealed that they were already one step ahead of him.

William, our plan is to have an online forum. And we hope you will be coming on board with Afronet.

Really? Why?

Well, we need someone that knows how this stuff works. When we start having our chats, we need to have them produced. And someone needs to be able to sit at a laptop on demand almost anywhere in the country. So if we bring on Joe Louis the boxer, and he's only got ten minutes, and we find out about it Wednesday, we've got to have that laptop in his hands with someone on the phone to support him. And that's you.

Oh, okay.

Come on down to D.C. We'll fly you down and talk to you about our plans.

And that's when I first heard that they were basically going to build a huge version of the Afronet. That is what they were doing.

Jenkins and his *American Visions* team had planned to build a series of chat rooms. Through a computer network like Afronet, they could build an environment where the public could chat directly, and in real time, to black celebrities and other public figures.

William, we want you to invite every other Afronetter to talk. I think we may piss some people off, so we'd like you to go and let them know that we want every one of them involved. But we're going to leave it up to you to organize it. And then we'll host it and make them feel good and we'll explain to everybody, give everyone an equal opportunity to get hired, to have a production room. We want to build a really big inclusive tent.

Oh, I see. Uh huh.

They [American Visions] *didn't really know at that time whether they would be on CompuServe or where. They just knew they were going to build this thing. So I jumped at the chance, because to me that was very exciting.*

So I went there and I tried to talk to several Afronetters. And that's when I got the pushback. Some showed up though. I think I invited thirty, and I think maybe something like seven or eight showed up. And it was a very productive meeting.

William explained the concept to his fellow Afronetters. They would run interactive, text-based chat "forums" on their boards, as all of them had routinely done. Only now, those forums would feature black celebrities and popular cultural personalities. They would do it on more sophisticated hardware and the latest networking software. Only, someone else—not those Afronetters—would own it.

Jenkins's instincts were right. Some of these folks got pissed off.

When they did, they directed their criticism and anger at William, not Jenkins.

Murrell, you're a turncoat. You're a sellout.
What do you mean sell out? We just built you a bigger platform!
But you won't be independent. You'll be taken over.
Well, kinda. But that's beside the point. Don't we want more people online?
Don't we want to reach more people?

William understood where they were coming from, even though he was prepared to move forward.

Several of them just stood right where they were and boycotted this new thing. And the rest came on board just fine. They felt like, and they just said, "Why don't they just take us as a group to running the whole thing for them? Why do they still have to maintain control?" Because again, the details weren't being disclosed. Again, we didn't know they were going on CompuServe. But when you're on CompuServe, you can't have fifty people joining the thing. You've got to have one entity sign the contract. You know.

But I think that's the psychology of it all. They [American Visions] kind of liked us because we had something going on. It actually was an operational network. It was valuable. It was sustainable. And then here comes a big, better-financed organization, looking at what they're doing, and saying, "All right, we just want to duplicate that or make it bigger." And I think that's what some SYSOPS and I felt like. But I'm used to integration and upgrades and all that stuff.

So then I go a different route. I say, "Well, we're moving from Platform 1 to Platform 2. That's what we're doing here." You know? Now keep in mind that some of these people, some of these SYSOPS were homemakers. Some were admin assistants at law firms. A lot of them were firemen. Some were involved in the ministry. But there was a lot of them that were firemen. But I say that like there's not been so many involved from the fire department, because they kind of stick together and know each other. The Afronet as a network probably wouldn't have grown as fast as it did without them. Because they make good money. So they have money to blow on these. This stuff wasn't cheap.

You know, you had to buy a computer. You had to dedicate a phone line in your house. You had to buy all these modems. You had to keep it all running. It was expensive. And no one's paying you. You know, you let people on for free. They're not paying you for it.

It was about growth. It wasn't so much about the money. It was more like who else can run a host? Who do you know that can run a host that would want to join the Afronet? Because, see, if I had permission, you can put up host and you're still not on it. Someone has to connect the chain, so to speak, or the command center, that actually puts you into the loop.

Anyway, we had this meeting. And one week later they [American Visions] announced the gathering of eagles, with the Congressional Black Caucus. And that's when it was officially released. The GoAfro thing was officially announced to the world. And at that point we had already hired me. They had already signed a contract saying, "You're hired." Get your buddies to help you and, you know, whatever you need and you'll have a budget. You can't hire everybody but, you know, you pick and choose.

Again, Jenkins used his magazine to announce the upcoming gathering. He dubbed it *A Gathering of Eagles.*

Fortunately, American Visions can now accommodate all of those willing to take the adventure to the treasure houses of the world through AV Cyberscape. This new electronic universe includes our African-American Culture and Arts Forum, available on-line through CompuServe; the worldwide marvels of Internet telecommunications, to which we can facilitate connection; and our comparatively low-cost capability to publish 300,000-page warehouses of information, sound and images on a single interactive compact disk. The beauty of these tools is that they allow any of us to wander anonymously around the globe, mixing freely in the potpourri of world civilizations at a fraction of the cost required for physical travel.

Jenkins had recognized the same needs and the same opportunities that not long before led Kamal to start AfroLink Software. Jenkins marshaled his influence as publisher of *American Visions* magazine, and his long connections to Washington's black political elite, to forge a new partnership with CompuServe. The company was the first commercial Internet service provider. Its roots stretched back to its computer time-sharing business back in 1969. The summer 1994 partnership with *American Visions* was as much about commerce as it was forging community, or claiming political territory. Nevertheless, Jenkins advanced all three interests in this same missive. It appeared on the first page of the June 1 edition.

Before Congressional Black Caucus Chairman Kweisi Mfume even announced it to his own members—which included thirty-nine members of the US House of Representatives and one member of the US Senate—Jenkins again beckoned black America to Washington.

> In mid-September (rather than November, as previously planned), AV will collaborate with the Congressional Black Caucus Foundation (CBCF) to host a Computer and Telecommunications Seminar and Expo to explore fully the potential of such implements in the context of leadership development. This will be an extravaganza of information and insights for intellectual empowerment unlike any prior national convention known to our community. Your participation in this information diffusion should compete favorably with any prior demands on your time during these three critical days. Plan now to come to Washington, D.C., for this nontraditional extension to the September CBCF Legislative Weekend. It promises to be a gathering of eagles and to offer revolutionary lessons in new ways to fly. Both you and the next generation will be thankful for the investment of attendance in this first celebrity event.

* * *

The CBCF annual event had been a Washington, DC, tradition. The weekend routinely drew a wide cross-section of black America numbering in the thousands. Equal parts celebration, fundraising, education, and activism fueled the event. The program featured what the caucus called "braintrusts" and issue forums. They designed each to inform and bring attendees up to speed about the latest social, economic, and political concerns facing black America.

Jenkins seized this unprecedented opportunity to foist technology, computers, and the new, so-called information superhighway to the very top of black America's political agenda. The conference had already emblazoned

Jenkins's interests in technology onto the annual event's agenda. The CBCF explained what that agenda entailed in the press release they sent just ahead of the event.

CONGRESSIONAL BLACK CAUCUS FOUNDATION SPONSORS
24TH ANNUAL LEGISLATIVE CONFERENCE

"Embracing Our Youth for a New Tomorrow"
Washington, D.C., September 1, 1994—The Congressional Black Caucus Foundation (CBCF) will convene its 24th Annual Legislative Conference (ALC) on September 13–17, 1994 at the Washington Convention Center....

The Annual Legislative Conference is a five day event which includes a series of issue forums, workshops and congressional braintrusts. There are also four fund-raising events which support the Foundation's educational programs and its mission to broaden and elevate the political influence of African Americans....

The information superhighway revolution will take center stage at the 24th Annual Legislative Conference on Wednesday, September 14, 1994 during an all-day conference on "African Americans in the Telecommunications Age." The conference will be held at the Renaissance Hotel located across the street from the Convention Center at 999 9th Street, N.W.

The Conference will examine the impact of technology and the information superhighway on the African American Community. U.S. Secretary of Commerce, Ronald Brown is the keynote speaker for the opening plenary session. Congresswoman Cardiss Collins (IL-7), Chairwoman of the Congressional Black Caucus Foundation is the keynote speaker for the conference luncheon.

Workshops will include discussions on business and job opportunities in the telecommunications industry; ways to ensure that African Americans are equipped with the critical high-tech educational skills needed to access and effectively compete in an information-based economy; technological resources for the nonprofit community; family life and work in the age of technology; industry regulations to avoid further divisions between the information rich and the information poor; the development of "inclusive" software; and African American entrepreneurs building and planning the information superhighway. The conference will conclude with a CEO Roundtable in which African American corporate executives will discuss how to effectively access the telecommunications industries and the future outlook for the Black Community.

The ALC conference took place just as Faith Edwards and Kelvin Dickerson had described in the CBCF press release announcing the event. The Caucus, Jenkins, braintrust participants—all of them had high hopes for the event. But they never predicted the kinds of success the event flaunted.

According to Missouri congressman William Lacey Clay, the event drew 30,000 attendees, with 152 media outlets from 21 states and Great Britain covering the proceedings. The program spotlighted more braintrusts and workshops than it had ever before featured. Still, technology, computers, and the Internet remained front and center from the beginning to the end of the conference. Sponsors were plentiful. The National Society of Black Engineers organized a two-hour reception for the "African Americans in the Telecommunications Age" conference. NASA orchestrated a high-profile hiring expose at the event's procurement fair.

But not everyone who participated celebrated the moment, or the event's outcome. Ken, Kamal, and William had been seated side by side on the panels "Toward a More Inclusive Software" and "African Americans and Telecommunications: By-Standers or Participants in the Next Electronic Frontier?" Brooklyn, New York, Congressman congressman Ed Towns had presided over those gatherings. As much as he wanted it to be otherwise, the event hit Kamal like a punch to the gut.

I remember those meetings. I remember a conversation with [Detroit, Michigan, congressman] *John Conyers. It's funny ... it was one of those kind of things where you have folks from everywhere—they say we're not monolithic— and there are times I want to believe that and there's times we should be mono-lithic. You had all flavors and variations.*

When I started Afrolink, it was not driven by money. . . . It's about purpose and meaning. I wasn't looking for anything in particular. I wasn't ivy-league educated. There were many brothers and sisters there from the ivy leagues. I had disdain with corporate America, and many there were chanting to corporate America. And so, I didn't go there with any preconditions or any expectations. I was just going there hopeful that there would be some alignment, some consensus—of an agenda, a plan, and execution. I met many people over the years…and I heard a lot of rhetoric. From that, what do we do now? What's our plan? What's needed? When do we start? Let's execute, let's build. Going to CBC in 1994, the expectation that we wouldn't seek to insert and become part of what's there, and create something new. We shouldn't all be standing in line to work at IBM. We need to build IBM.

Kamal went to Washington hoping to plan what Jenkins had promised: a revolution. What he saw all around him was not even close. It wasn't even disruptive. If nothing else, the weekend had impressed on him a lingering question. Was a revolution in technology even possible? One that would put technology in the hands of black people to further our interests first and foremost? Kamal stood to make millions selling black software. But for him, commerce was a means to an end: community uplift, economic empowerment, amassing intellectual property, black influence and control. He sought those ends. But he feared that the future of black America's tech revolution was already turning us into slaves to commerce. Making community concerns second place, if they placed at all.

Regardless, Jenkins coordinated with his band of willing Afronetters, members of Congress, organizations interested in furthering science and technology. Together they took advantage of this unprecedented opportunity. It soon materialized into an unprecedented commercial venture that thrust Jenkins, his new platform GoAfro, and CompuServe itself into a fierce competition.

Chapter Eight

THE BATTLE FOR BLACK CYBERSPACE

Beginning on April 12, 1861, America engaged in a great civil war. On January 1, 1863, President Abraham Lincoln's Emancipation Proclamation gathered legal force. For two years, five months, and nineteen days thereafter, nothing changed for many slaves. Then, Major General George Granger and Union troops arrived in Galveston, Texas. There he read the proclamation:

> The people of Texas are informed that in accordance with a Proclamation from the Executive of the United States, all slaves are free. This involves an absolute equality of rights and rights of property between former masters and slaves, and the connection heretofore existing between them becomes that between employer and free laborer.[1]

No slave was free until all slaves were free. And so black people commemorated the day. They called it Juneteenth. The day both ended slavery and ushered in a brief period of Reconstruction.

* * *

One hundred thirty years later to the day, David Ellington and Malcolm CasSelle symbolically assumed General Granger's role. Their announcement was as revolutionary as that historic moment when that last slave received word that she was free.

It all started earlier in 1994. Timothy Jenkins worked on the sly back east to launch GoAfro. Meanwhile, Malcolm beckoned David back down to The Dungeon. There, Malcolm gave David a glimpse of the future.

That's where Jerry Yang was, ya know, and people like that. And I showed him Yahoo! for the first time. It was the early days, and Jerry was one of my classmates. We were really at the frontier. David was excited by it. Again, coming from the perspective of someone who knew nothing about computers. Coming from the world of a lawyer, who used a PC for word processing. We were accessing the Web. The Internet. So imagine going from a computer that was not connected to anything, to a computer that is connected to everything. He was like, "What? I mean, this could do so many things." You can see all of this information. You can find all this stuff. David was fascinated by it. To me, I was like, well, yeah. . . . Of course. I mean, it wasn't novel to me. It was the way things were. But I said it again—the future is about connected computers.

David was savvy enough to see the future with Malcolm—even if he didn't fully understand it. But, he understood his clients. He understood that the arc of history did not always bend toward justice.

I mean, I didn't know anything about money. I was tired of seeing my black entertainment clients get burned in LA in entertainment. I saw this new medium as a new way that not only could my clients and potential future clients be a part of this distribution channel, this platform that was coming called the Internet. Because, remember, I had secret insight. My partner was at Stanford in the computer science lab in the dungeons.

I saw the kids were in on this thing, coming up with content or using it to do research. I had used LexisNexis in law school in the eighties. You're doing your dial-in, you're looking up a case and you type in the case code and the name and then you have to read cases related to the case. I could learn case law, but it was all text. What these guys were showing me was this new medium, because this thing called the World Wide Web came along, WWW, and that's what gives you audio, graphics, sound, whatever. So anyway, he and I then decided this is really interesting, maybe there's a way that you and I could work together.

* * *

America Online (AOL) started in 1983 as a gaming company, Control Video Corporation. When it reorganized toward the close of the 1980s as America

Online, Steve Case, one of the company's founders, became its chief executive officer. Case quickly orchestrated a series of financial and personnel decisions designed to lift AOL high atop the Internet service provider market.

By 1994, the company was on track. But Case saw that AOL needed more aggressive marketing. It needed to capitalize on its existing market to build new ones. Case did not wait to develop the aggressive, guerilla marketing prowess that AOL needed. He bought it. In 1994, AOL acquired Redgate Communications Corporation. The computer marketing firm published magazines such as the *Amiga Buyer's Guide*. As its name suggests, the magazine/marketing tool reviewed hardware and software related to Commodore's Amiga computer.

Even more than the company itself, Case wanted Redgate's president, Ted Leonsis, who had made his first millions—forty to be exact—when he was just twenty-five. Leonsis did not believe the adage that "content is king." At least not in the same way that Lee Bailey had embraced the principle. Sure, he believed that you need great content to attract audiences. But Leonsis thought community was marketing's strongest foundation.

Community and providing context is what's important. We will always remain member-centric, not IP-[information provider] centric.

Leonsis started looking for people who could bring other people together. And he came up with a plan to find them. As 1994 drew to a close, Leonsis announced a bold first strategy to grow AOL's subscription-based Internet portal. The goal: get a million and more Americans online. He looked for what he called "infopreneurs." He did not imagine run-of-the-mill content producers. The magazine world he came from and the media business in general were awash with such mere mortals. Leonsis searched for people who could bring other people—by the millions—to the online community AOL was beginning to swiftly build. Leonsis named his venture The Greenhouse. Its plan was simple. Find a handful of companies, led by people who could help AOL accomplish its goal; provide them capital; take a minority share of the company in exchange; throw AOL's recently acquired marketing weight behind them; then sit back and watch the dream unfold.

* * *

David explained.

And then literally in '94, I was dating a woman who was working for a multimedia firm—so this is actually all relevant—who was working for a multimedia *firm, a brand-new thing called a multimedia interactive agency, called Redgate Communications, here in the Bay Area. It was run by a man named Ted Leonsis. There was this other little company in Washington, DC, called America Online, and it had a $400 million market cap and it was trying to grow. And there were all these other online services called Prodigy, CompuServe, Genie from General Electric, Apple's E-World, and they were competing. And what they started out with, like everyone else in the entertainment business, everyone was focusing on CD-ROMs. But what Ted saw after publishing a few CD-ROMs with catalogs on it, he saw that as you put it into the computer, it could also dial the modem and connect to an online service.*

He showed that to Steve Case. Steve Case then bought Redgate communications, and then Ted said, and this is in 1994, Ted said, "What I want to do is develop some infopreneurs," as the new president of America Online in 1994. "I wanted to identify different types of pure online content people. And I'd like it to be [as] diverse as possible." My girlfriend, who worked for Ted, is now an employee of America Online because Redgate got bought by America Online.

She said to Ted, "Hey, I have a boyfriend who would like to start something black and black entertainment–related. What do you think about that?" Ted said, "I love it. Get me a business plan."

David was the approaching-middle-age lawyer. Malcolm was the young geek. David took the lead. His vision was dead on. But Malcolm was there to remind David that his execution—his proposal—lacked, well . . . a kind of technological charm.

We suddenly realized that the idea of a network of black culture was an opportunity.

The two tossed around potential names for the venture that began exploding in their minds. *Afronet* was first out of the gate. They ruled it out. They discovered that a company with that name already existed—selling hairnets! Not to mention, Malcolm also pointed out that an online service named Afronet also already existed. Malcolm suggested *CyberBlack*. David squashed it. *Too hard*, he said.

I could have easily gone down the path of trying to be this blacker-than-black service. But then, I had to say, hold it—we're about to enter the 21st Century. And it's going to be about communication. It's about creating a place for people to talk, debate, and have fun. To me the business model of the next century is about inclusion.[2]

Then it happened, in tandem:

Net...

Noir.

David typed up a document that was the basic idea. Like a one-pager. Or two-pager. And it was all in like Courier font. It looked like it was done on an actual typewriter! I think it was actually a word processor. And I looked at it, and I was like, w-e-ll, yeah, okay. It was from a lawyer's perspective—ya know, bullet items and, it was like an outline. And I said, "I don't think it actually works that way, but I get your point." So that's how it started. We built a team, got the business plan together.

Then, Malcolm and David walked into Redgate, sat in front of Leonsis, a young, now-larger-than-life entrepreneur who sat on top of the online world. Not fazed by celebrity, wealth, or the fact that Leonsis stood to make or break their dream, David and Malcolm made their pitch.

NetNoir. What was it? What could it be? Why should it exist? How would they do it, and what made them think they could be successful? David and Malcolm laid it all out for Leonsis.

> *This is a cybergateway into Afrocentric Culture.*[3]
>
> *We believe our content will be very appealing to universities, especially the 103 historically black colleges and universities.*[4]
>
> *NetNoir will be a study in synergy. Our advertisers, our content providers and our merchandisers will all come together and mutually promote each other while focusing on Afrocentric Culture.*[5]
>
> *This is a cultural imperative. The service's software will drive the purchase of hardware by people who are interested in Afrocentric information and opportunities online, but aren't already involved with technology.*[6]
>
> *It is global in scope, including Afro-Caribbean, Afro-Latin, Afro-European, Continental Africa as well as African American. Anyone who has an interest in Afrocentric culture can come in and participate. You don't have to be black.*
>
> *If we create software, that is, faces, entertainers, literature, you know, academic pursuits, that look and feel like things that you're comfortable with, there's a reason for you to get online.*[7]

Malcolm and David walked away from the meeting confident. But they were realists. They knew theirs was a shot in the dark. They had applied to Leonsis's Greenhouse program, just like seventeen hundred other individuals, teams, and companies vying for America Online's money, marketing, and home within its online portal. There were seventeen hundred other people also convinced that their idea could work.

David and Malcolm had been part of many different kinds of pitch meetings across Silicon Valley. Those meetings didn't so much put them on the defensive, but they did put a chip or two on each of their shoulders. They were usually the only two black people in the room, and had to fight to get what seemed to come easily to their white competitors. But those meetings had also helped David and Malcolm realize that they worked for more than just their own personal financial hopes and dreams. NetNoir, they both firmly believed, would not just be good for them (though, to be clear, they both planned to get filthy rich). They also believed their success would translate to something significant and positive for black people. They willingly shouldered that burden. If we don't succeed, they don't succeed.

I was going to literally—Malcolm and [me]—for '93 and '94, we must have gone to eight or nine different conferences. We were always the only black guys in the room. And we were sitting there watching, and at that time it was only white guys with pocket protectors. It was truly geeks. No women at all, maybe one or two, and they're super-geeky, too. And it was just really like four or five hundred people in a room. I can remember one conference in particular, and no one was talking about anything black. They were always focusing on the Grateful Dead, the Rolling Stones, and then I heard Prince was doing something. . . . And it was like, "Oh, that's nice." I was like, Oh, no, no, no. I want all of our stuff to be on here. You're not going to build this new medium and cut us out.

Determination kept them going while they put the pieces of their company together. But determination alone wasn't going to build the company. They needed AOL. They needed Ted.

* * *

With these kinds of endeavors, you bet the jockeys and not the horse they're riding. They both showed up for the pitch meeting and they were able to articulate with a gleam in their eye exactly what they're going to do. Having a gateway to the entirety of Afrocentric culture was appealing. More than that, they were more appealing as a team. They could have come in and sold us anything. I knew in the first fifteen minutes I was going to do this deal.[8]

* * *

Again, David provided details.

We were the first company funded in this thing called the Greenhouse Program that was launched by America Online. They did it as a marketing tool, as a vehicle and as, basically, it's an exercise in vision. Because what Ted got is that online content would be compelling and would drive people to use this new technology called online service, because everyone knew that CD-ROMs were purely a hybrid technology. It was going to lead to something bigger. And that bigger thing became known as the Internet.

Remember, the commercial aspect of the Internet—the Internet has been around for thirty years for university professors to talk to the DoD, the Department of Defense—but the commercial aspect—there has to be sexy and compelling content. And Ted had the vision and convinced Steve Case, the chairman of AOL, that that's what you need to do. We need to have original content and we need to identify infopreneurs and that's what he did. And this black company was the very first company funded in that vision.

To say that David and Malcolm were ecstatic would be a severe understatement. They knew they deserved the multimillion-dollar vote of support. But it's a whole different thing when the man holding the bag of money realizes it too.

Though, on a more practical level, Malcolm and David were excited because they, personally, *really* needed the cash!

David and I were starting up the company and basically we were broke. We had quit our jobs. We shook hands with AOL, but we were still negotiating the terms of our contract with AOL . . . we knew that we had to get moving in order to get a jump on what we were trying to do, so that we could be the first, we didn't have much time on our hands to mess around. We had asked AOL for an advance on our financing and for it to be rolled over into our equity . . . and they agreed to send us a check for $30,000.

Well, the first thing they did was FedEx the check to me in San Francisco. Mind you, I was in San Francisco and David was in Los Angeles. However, the check was made out to both Malcolm CasSelle and David Ellington. But of course at the time we did not even have a joint account together. We weren't even in the same city. David would have to come up to San Francisco or I would have to go down to Los Angeles.

We had no money. So, I borrowed money from a friend to get a one-way ticket to Los Angeles. We drove to Beverly Hills where David had a bank account at the Bank of America, and tried to deposit and cash the check. We needed access to this money. They took the check and said, "I'm sorry sir, we can't cash a check for *this amount. However, since you have been such a good customer, we can advance you $1000. It will take two weeks to clear...*"

We said, "*No, you don't understand. We need this check now!*" So the teller walks away and comes back and then says, well okay, even though you don't have two cents in your account to your name, we will advance you $5000! But then, the teller goes away and comes back and says, "*I'm sorry, but there's insufficient funds.*"

We couldn't believe it. AOL gave us a rubber check! Here I am in Los Angeles on my one-way ticket, and we have this piece of paper for this check with all these zeros, and it's useless!

So we call the AOL office and it turns out that the person we did the deal with, Ted Leonsis, is out on vacation. And so we explain the story to his assistant, and she ends up calling him on this mobile phone in Florida where he's on a boat. And we said, "*Ted, they just told us there are insufficient funds in this account.*" There was a long pause and then he screams, "*What do they mean there's insufficient funds! There is $60 million in that account! Okay, I'll call you back.*" And it ends out working out okay. I got back up to San Francisco and we get the money.[9]

Once they got their money, the NetNoir team went to work. As they did, they honed their vision. That vision was now joined in wedlock to AOL's vision.

None of us knew where the hell this was going or where it was leading. But Ted had the vision to say, "Wait a minute. This is basically a new medium." And it's the new television. There's got to be television programming. And it's not going to be just for geeks. It's going to be for the masses. And of course, Steve Case's background came from Procter & Gamble, so he obviously was always focusing on consumers, and AOL was that. But they first started with financial channels and really clunky and you click on things, and lots of text oriented, not video. I mean, we take it all for granted now, what we can do with our mobile devices, let alone what we could do with our PCs back then.

He was like, "Oh no, if there's anything about black culture, it is compelling and crossover. You guys go build that. Make that happen." He's focusing on crossover. I'm focusing on making sure my culture is a part of or is directly a dead-center part of this new media. I ain't going to lie, I had no idea where this was going. Ted, none of us. None of us knew. I know Mark. I know Chris Larson of E-Loan and Larry Page of Google when they were just getting funded. None of us knew that it was going to be this freaking big and this revolutionary.

We thought it was going to be big, but no one knows. We were just trying to build something. It was cool. We really did drink the Kool-Aid. It's going to change the world. It's going to democratize the entire planet when people have access. More ways for people to vote that couldn't vote before.

When I saw how big it was and when I saw Ted was taking me seriously, I was like—Oh. I was determined to make sure that my culture, that black culture was a part of this new media. But I was just driven by I wanted to make sure that black culture was on this medium. So I had meetings with Ebony. Black Entertainment Television *tried to buy me, Bob Johnson. I was going around—* Essence. *Everyone was trying to watch my shit or go around me and go to AOL. And Ted, to his credit, pushed everyone back.*

To his credit, rather than alienating them, David, as NetNoir's CEO, struck deals and brought them into the fold. He made allies and content partners out of the legacy black media outlets. The outlets would provide NetNoir content. NetNoir would provide what everybody needed: visibility.

* * *

It was Juneteenth, 1995. The world's black diaspora had something more to celebrate than their century-ago release from bondage. Reporting for the Associated Press, Elizabeth Weise wrote, *It took until June 19, 1865 for Texas slaves to find out the Emancipation Proclamation had freed them two years earlier, an event the black community celebrates as Juneteenth. Now a new Afrocentric on-line service seeks to ensure that blacks never again have to wait that long to learn about their culture, and founders chose Juneteenth's 130th anniversary to launch NetNoir Online.*[10]

David, Malcolm, and the NetNoir team used one of its primary services, a live chat room, to announce its presence to the online world. And, they did it using what would be one of NetNoir's signature content features. A celebrity live chat, or what they and others in the business called a forum. On that day, testifying to the moment's historical magnitude, David and Malcolm had invited Larry Irving, head of the US Department of Commerce's National Telecommunications and Information Administration. He leveraged the moment to highlight what was most on his mind. It was less about NetNoir itself, more about encouraging more blacks to get online. NetNoir's announcement chat forum was the last time that David and Larry Irving would see themselves as allies. They ventured in separate directions to build different sections of the new information superhighway. David was an entrepreneur. Larry was a policymaker. David had one aim—make money for himself, his company, and his investors. Larry concerned himself with how this new technology might positively or negatively affect all black people, and those without the financial means to find an onramp to the information superhighway.

But not everyone shared David's, Malcolm's, and AOL's optimism for NetNoir. After NetNoir's Juneteenth launch, skeptical media analysts, financial speculators, stock prognosticators, and advertising industry writers alike lined up to throw shade. Could NetNoir single-handedly integrate the whitest, most expansive, and potentially most lucrative new neighborhood on earth—cyberspace? People who questioned NetNoir's ability to live up to the hype that AOL had helped to manufacture throughout the spring of 1994 didn't necessarily speak out of turn. Larry Irving's department at Commerce had already leaked the findings from its landmark study. Irving had commissioned the study in collaboration with the US Census Bureau, and it showed that blacks were far behind other computer and Internet-connected users, with 20 percent more whites owning computers than blacks.[11]

Malcolm and David had built it. And to hell with the early Web's demographics. Fuck that study. David saw through it all. Nobody really knew at the time who was online and who was not. Irving's study was already too behind the times to properly assess cyberspace's market realities. Data for the study came from the 1990 census, supplemented by a specific computer study early in 1994. And the study's conclusions were based primarily on measuring computer and computer modem home ownership. But computer ownership wasn't everything. Schools have computers. So do libraries, churches, community centers, friends, and universities—all the places that black folk congregate and share knowledge and resources. Not to mention computers, equipped with all the necessary tools to get online, had invaded the workplace.

David and Malcolm knew their people would come. They knew that whether it was in computers or online services or any other material good— we had become a market from whom many could profit. David and Malcolm knew that they would make money from NetNoir. They hoped they would be role models for future entrepreneurs. They had built and begun to develop the space to make this happen, a platform from which other black people could launch their ventures.

* * *

Figure 8.1. Malcolm CasSelle, Cofounder of NetNoir. Courtesy of Malcolm CasSelle.

Farai Chideya had left Harvard, gone to work for *Newsweek*, then parlayed her work into freelancing. She was the first to take advantage of NetNoir's new platform to build something new—this time for journalism. It was also new for black people and those across the world who lived, understood, and recognized the ways that race, gender, sexuality, religion, and other factures shaped human experience.

Questioning the status quo was a family tradition.[12] Farai Chideya's mother had raised her daughter alone, oscillating between careers as a journalist and teacher. Farai arrived at such a question when she walked into Harvard Square in 1986. Sure, one part of that question focused inward. *Am I as sophisticated as all these kids around me?* she wondered quietly. But more important, as part of both Harvard's black and female minority, her presence at Harvard questioned the institution's business as usual. Her poignant and pointing prose challenged the status quo in the columns and stories she wrote for the *Independent*. The paper itself had begun as an alternative to the news presented by Harvard's official newspaper, the *Crimson*. Farai challenged that status quo again when she graduated with distinction in 1990: magna cum laude, with a degree in English literature.

Harvard was far enough ahead of the curve to have a computer lab on campus. Farai lived just a stone's throw away from the place where wizards still stayed up late,[13] but she never darkened its doors. As the decade of the 1990s threw open its gates, her concerns couldn't have been further away. She walked away from Harvard and began her career as a journalist at *Newsweek*. She continued to cultivate her voice and she had a platform, but it belonged to *Newsweek*. Soon she would find another; one of her own.

In the early 1990s, Farai had been invited to join an online community called the WELL (Whole Earth 'lectronic Link), a BBS launched by a group of hippie technologists in the late 1980s and early 1990s. Farai was one of few non-white folks in that community, but she saw the power the platform provided.

It grew into a cultural juggernaut where you had all these journalists from across the country. I remember reading Sinbad . . . he was on the WELL and he'd be online chatting. The bulletin boards got political, as people talked about everything. With WELL, anyone can create a sub-topic. People could start conferences and there were also closed conferences that were membership only, dealing with specific issues. It was like friends meeting in a coffee shop.

Farai was freelancing. Websites proliferated, but she was interested in building something that was much more interactive.

What was interesting to me was bulletin boards where there was active exchange. Then, you see blogs, and I was like oh I could do this, so I did.

By late 1995 Farai had connected with Omar Wasow, a young, black, digital upstart who had built a BBS called New York Online from his Brooklyn home. That community helped to birth what became PopandPolitics.com. PopandPolitics was one of the earliest blogs ever produced.

I saw PopandPolitics as a place where I could be free and say what I wanted without editorial oversight, which again is good and bad. With solo blogging . . . most successful bloggers, even ones who are solo bloggers, have someone editing them. I enjoyed the freedom. Newsweek *had rules about how much voice you could put into a particular piece.*

In her words, there was something going on—a new way of producing and distributing content. A new way of trying to produce targeted news for both public edification and journalistic profit.

Um, is it just me or is there something going on here? Well, both. Something's definitely going on—one of the hottest, smartest websites targeted at young

Americans—and then there's me and the rest of the crew that brings you "Pop & Politics." I'm CNN political analyst Farai Chideya, and the rest of my team includes guest writers and the web design team of New York Online.

With an incredible reader response to elements including an interactive quiz, an advice column, and numerous new essays, we've already built a strong following. Now we're actively looking for sponsors to help us continue and expand one of the hottest sites on the net. Contact New York Online President Omar Wasow (omar@nyo.com) for rates and information.

Farai capitalized on the blog's visibility. Then she doubled her efforts. The same year, she authored her award-winning book, *Don't Believe the Hype: Fighting Cultural Misinformation About African Americans.* She used PopandPolitics, and NetNoir helped to catapult the book to award-winning, multiple print-run status.

* * *

Figure 8.2. Farai Chideya, Founder of PopandPolitics. com. Courtesy of Farai Chideya.

Like Farai, Lee Bailey's first foray into the new online environment was connected to the work he'd been doing his whole professional career. It began when he expanded *RadioScope*. He had created an email newsletter called the *Electronic Urban Report*. At the same time, NetNoir soon encountered a serious challenge. The black masses that arrived at NetNoir directly—through other locations on AOL, or the open Web—voraciously consumed content. They needed to be fed literally 24/7 with stuff that was relevant and new.

NetNoir was like the black component of AOL. They needed content, so they reached out to me. They were aware of me and RadioScope, and they thought I'd be interested in providing them with entertainment content. Of course I was. It was perfect timing. I was interested in the Internet and being connected with it. They needed content. It was a perfect marriage.

AOL was this closed wall garden, an Internet within Internet. The content was behind a wall. You had to be an AOL subscriber to get to their content. That was cool but I wanted to reach an even wider audience. Mike Blackstock, a Boston-based web host (globaldrum.com) found out about the newsletter and contacted him [David]. Mike would reproduce the newsletter on his website. At the time, the newsletter had complete stories.

We didn't have a website per se, but we had a newsletter and an email address. He contacted me and I thought that sounded like a good idea. At the same time new stories were set up on their own page. He wasn't that sophisticated and I wasn't either.

They repurposed the entire newsletter, with ten to fifteen stories per publication. It was one, incredibly long page of scrolling text that was primarily all text. He gave us distribution via website where people could find it online.

We found advertisers on our own. I never even thought about charging either NetNoir or GlobalDrum or anyone for the content. That came later. At that point it wasn't really a consideration.

It was a new medium, but NetNoir was playing in Lee's world: the content business, and the advertising and marketing business. Lee willingly distributed his content to NetNoir and others for free. Not to worry. Lee knew perfectly well he had to turn those eyeballs into cash.

* * *

By 1995, competition—sometimes tension—intensified. Some folks came to the game to make money, pure and simple. Others knew they had to make money, but they chased it in service of a larger, more community-, educational-, or activist-focused goal. For a short time, GoAfro made a run at NetNoir, with William at the helm.

When NetNoir was released, it was the first time someone like AOL spent money on anything like this. So they invested a half a million dollars into NetNoir. AOL themselves did it. They wanted this to be a success. So USA Today covered it. Barbara Reynolds wrote the article on it. And Charlayne Hunter-Gault was the star. In other words, when it opened, it featured Charlayne Hunter-Gault, which at that time was like a Katie Couric kind of person.

And so that was the draw for them, coming out of San Francisco. And it was being managed by some pretty cool black people who were very hip and sophisticated. And it was clearly designed to attract—in fact, Barbara Reynolds's words were—the urban intelligentsia, the black intelligentsia. It was designed to attract that young professional, you know, black crowd of that eighteen to thirty-four age group category. And the content they had was like lifestyle stuff, nothing serious.

Our project on CompuServe was created by PhDs, academic historians. They represented their magazine called American Visions. *It was the official magazine of the Afro-American Museum Association, closely tied to the Smithsonian. And so it had already had six years of publication and we just focused on that stuff. So if you wanted an in-depth treatment on the history of black/white cooks, we had that material. And that was the difference between our forum on CompuServe and their forum on America Online.*

So we were competing against each other. You know, CompuServe had us to brag about and AOL had NetNoir. CompuServe was very strict. It was like being in Catholic Church. AOL had no rules. If you wanted to curse, you could curse. And so people did it. People looked out for themselves, and they preferred the more open environment of AOL. So the same person probably was on both services, because it just depended on what they came to the service for. But ultimately the glue, or what I call stickiness, was always the socialized chats.

I hosted celebrity events. I had celebrities on there and we hosted those things. It's like, you know, you have a big interview situation with like a big-name person and everybody shows up for that. We had that. We brought that. But people came every day not for that. They really came to, like, chat with the girl that's cute and just put up a picture of herself in Los Angeles. And maybe you two will get together. All that was going on the side, and that was exactly what brought people back over and over, to the point where some people became addicted. I mean I remember women who would call me up, tell me that they just got cut off. They can't connect, can I help them, talk to so-and-so. I mean that was not unusual, man, I'm telling you. That was not unusual.

And so that was really the glue, that stickiness was that chat. Those chats kind of were private. You could be sitting in the forum with the forum content, what's needed to sponsor stuff, so to speak, while simultaneously you've got three open chat windows, talking to three different people at once. And that was really what kept people there, I think, in my opinion.

Both GoAfro and NetNoir had found that more than anything else, people flocked to their respective sites because they could connect with people whom they never had the opportunity to otherwise. All those lofty platitudes that David had mentioned about strengthening democracy, increasing minority participation in politics, and saving the world and all that shit. It paled in comparison to making a love connection. It doesn't mean that both platforms didn't try. It just meant that both began to see the realities of how and why everyday black people would want to use this new medium. And unfortunately for GoAfro, NetNoir and AOL were built to capitalize on this reality. CompuServe was not.

* * *

Some of the Vanguard had hitched their wagons to the two corporate ISP giants, AOL and CompuServe. But others tried to edge their way into the market on their own.

Back in 1994, Derrick had given Alou "Lou" Macalou and the BGSA's tech committee the green light to pursue building the UBP. At the same time he began to channel all his energy into his own brainchild. He called it *Knowledge Base.*

15 MAR 1994—INTRODUCTION / PURPOSE & SCOPE OF BUSINESS

To take full advantage of the information superhighway and all of its educational & business resources, it is becoming increasingly necessary for families and individuals to own (or at least be familiar with) personal computers. This is because the PC is the primary link between individual users and the vast network of services available via the Internet, which is the backbone of this superhighway.

What I shall describe here is a business plan that focuses on offering consulting/teaching/development services that will cater to a broad spectrum of individuals ranging from first-time computer buyers, to those who wish to computerize their small businesses, and to those who wish to become knowledgeable of information access tools available via the

Internet. To reiterate, my markets are: First-time computer buyers; Small business owners; Those interested in accessing the Internet.

These activities will be defined from my generic skill set, which currently features writing, researching & learning, and teaching. Immediate results could be obtained from the following activities (technical): freelance writing & consulting; technical; image processing & general DSP; computers in the home & small businesses; and (non-technical), negotiating the Ph.D. process; financial resources for graduate school; on-campus tutoring (mathematics & junior-senior-level Electrical Engineering courses).

Derrick proposed to counsel first-time computer buyers on their purchasing decisions by showing them what hardware and software best fit their needs, setting them up with intelligible instruction manuals, and providing them how-to books to help maximize their purchase. He would teach them how to maintain their systems, provide in-home software tutorials, and show them the real-world uses for popular software applications like Microsoft's Office suite. He thought he would offer the same services to small business owners. Some added services would be appropriate of course: consulting in business-related computer applications; setting up an office network; configuring email; and choosing an Internet service provider, like CompuServe or AOL. Maybe he could even offer a crash course in programming, using Unix tools such as vi, sed, awk, perl, yacc, and shell programming.

By 1995, Lou and others had begun building the UBP. Derrick still led Georgia Tech's BGSA. And he began using it as a vehicle to launch activities consistent with Knowledge Base's vision.

Beginning in 1995, now all of us could pay attention to our research in our day job. You know, the UBP really was our research. And it's 1995, so there's no major, even at Georgia Tech, no information technology. It's not mature enough yet for it to become part of the academy. So we're going on our own devices, and our advisors aren't really concerned because the advisors are walking through the labs to see those nerds on the computers, and that means research is happening. So they leave us alone, and we build it out and one day Lou tells me, "I want to make this into a company, do you mind?" I said, "Well . . . no. I don't mind, that's kind of what we do in this environment." The black student office area of the Society of Black Engineers is next to [the] grad student association, which was right next door to the African American Student Union.

And when you're all in the same office space like that, you're always doing entrepreneurial things, so there are a lot of small businesses that you can run out of those offices and build into larger enterprises. But the spirit is that we're all about trying to get together and build stuff. So I tell 'em, Lou, you go 'head and do it man. I wish you guys well because I'm trying to build something too, it's called Knowledge Base. Who knows, at some point maybe we join these, because the vision of Knowledge Base is to take advantage of who I've been all my life. I'm the guy that everybody comes and says, hey, can you show me how to do this? I said, okay, I can leverage that. I can definitely leverage that.

So here's what I envision. You guys go put this UBP together. Maybe I can get together with you and Knowledge Base, and we can do outreach. Like grass-roots outreach. There's a lot of people at this school where everybody's smart. But there's a lot of people at this school that know nothing about this. Imagine what's going on out there in the city, in the country, anywhere we go, we can introduce people to this stuff. And we can build our own office. It makes sense.

So fast-forward to the end of 1995. Lou calls me up and he says, "Hey man, I'm about to do this, and I think I'm gonna leave school. What are you gonna do?" I said, "Okay, well, I don't wanna leave school just yet . . . so if you're serious, let me know how you want me to fall." So he came back and he said, "Well, I think you're the only one among us engineers who knows anything about business. So can you build this into a business?"

And so, that's what I do. I show people how to do stuff. So sure, I can build this into a business but here's the thing. I'm the architect. Seldom heard, seldom seen. I'm like the glue. I'm going to build it, hold things together, then I'm gonna disappear. And hopefully what I put together will stay together. That's gonna have a lot to do with you, and who you bring in. Because I know you. I don't know the guys you're bringing in.

Derrick fulfilled his promise. Early in 1996 he began to ink a marketing plan—the blueprint for UBP's business enterprise. He called it BGS Infosystems.

EXECUTIVE SUMMARY
BGS Infosystems, Inc. Marketing Plan CONFIDENTIAL

BGS Infosystems Inc. ("BGSI," "The Company"), founded in January 1996 with two employees and currently headquartered in Atlanta, GA, is seeking initial equity financing of $99,980 (9998 shares of outstanding stock at $10 per share). This funding is desired to establish BGSI's business infrastructure by allowing the purchase of computer and office equipment, the leasing of office space and communication hardware, the retention of a lawyer, accountant, and independent contractors, and the payment of employee salaries. This figure will increase significantly once all financial projections have been completed.

This desired infrastructure will support BGSI's primary business activity—the development, maintenance, and marketing of an Internet directory accessible through the World Wide Web (WWW) called the **Universal Black Pages**™ (**UBP**). The UBP is a comprehensive directory of African-related Internet resources that has been under development since July 1994. The primary purpose of the UBP is to encourage development of African-related categories and topics which are not currently available via existing WWW pages, which will result in a greater WWW presence for members of the African diaspora. BGSI's effort to motivate development of WWW content and to maintain a comprehensive directory of that content will result in the following user benefits:

- Users enjoy the time they spend browsing the page—they are *edutained* (that is, they are educated and entertained).
- It can be used as an educational resource (in conjunction with traditional media like books, periodicals, and films) by students and teachers at all levels of education.
- It can be a vehicle for promoting use of the Web and the Internet among people of African descent.
- It will serve as a networking tool for people with African interests.
- It will offer a more global perspective of African peoples and cultures.

BGSI plans to produce revenue with the UBP by selling advertisements to companies who desire exposure to the UBP's niche audience—persons of African descent who range in age from 18 to 35. The UBP currently has 18 advertising spaces available. Leasing all 18 spaces to only one advertiser apiece can generate $45,000 in gross monthly revenue ($545,000 per year). BGSI plans to utilize current WWW programming technology to develop techniques to display multiple advertisements on the same page, similar to the way advertisements rotate on the scorer's table at professional and collegiate basketball games. This could significantly increase projected revenues.

BGSI has carefully studied and researched the commercialization of the Internet and WWW for the past six months. This research has indicated that there are currently several general Internet directories who are accepting advertisers. Some of them have acquired as many as 50 customers. These general directories are accessed on the order of millions of times per day and gross monthly revenues as high as $59,000 *per advertising customer*.

The UBP is a niche product currently accessed on the order of thousands of times per day. After researching other Internet directories' advertising rates, BGSI has scaled its advertising rate in accordance with its smaller audience. The Company's research indicates, though, that the UBP is currently the largest Internet directory of its type (i.e., a directory of African-related resources), and has been in development at least nine months longer than the nearest direct competitor. As of the writing of this plan, none of the UBP's direct competitors are accepting advertisers. The UBP plans to be first.

The core technical operation of the UBP is performed by internal software modules called *engines*. Among the engines that are either currently employed or under development in the UBP are the following:

- **Search Engine**. The search engine is proprietary software that allows a user to locate specific entries in the UBP database using keywords.
- **Survey Engine**. The survey engine is not proprietary software. This software was developed by Jim Pitkow, a graduate student in the Georgia Tech College of Computing. The survey engine employs features which validate answers before terminating a survey.
- **Registration Engine**. The registration engine is proprietary software which steps users through the process of registering information in the database. The engine actively checks the validity of information as it is submitted. Once registered, the data is stored in a database for further processing by the administration engine.
- **Page Builder I Engine**. Page Builder I is proprietary software which is still in development. It dynamically "builds" HTML-coded pages at regular intervals, incorporating the most recent changes to the UBP database.
- **Administrative Engine**. The administrative engine is used solely by the UBP staff to maintain the UBP database, change graphical user interface (GUI) formatting, maintain sponsorship, and perform other site administrative tasks.

The core of BGSI management is comprised of two African-American males, Alou Macalou and Derrick Brown (both of whom hold advanced engineering degrees from Georgia Tech), who together possess a strong fundamental knowledge of both the technological and business aspects of the Internet and WWW. BGSI's management core is also well-versed in emerging computer technology, and contains individuals who are strong leaders, effective communicators, and have a clear vision of how the Internet and WWW can positively impact the African diaspora.

BGSI has also assembled a support team that includes directors with over 25 years of marketing experience, advisors who are at the top of their respective academic fields, a highly skilled group of individual contractors, and sound professional support personnel. A comprehensive business plan with detailed financials will be available in late May 1996. BGSI management is pleased to discuss details of this business plan or its marketing plan, and to render demos of the UBP to serious investors.

Derrick did what he promised. He was the architect, and this thirty-eight-page business plan was his blueprint. It was Derrick's time to disappear. But circumstances just wouldn't let him. And that's what began to spell the beginning of the end, not just for Derrick, but for all of the Vanguard.

A battle for black cyberspace meant there would be winners and losers. Would the Vanguard determine their own fates? Or would someone else pick who won and who lost? The answer to these questions was about to become crystal clear.

* * *

Derrick had left Lou alone to run his own show. But Lou pulled Derrick into the mix, hoping to leverage his skills to launch UBP. Once he was in, it was hard for Derrick not to take umbrage with the direction in which he saw Lou and his team taking UBP.

Lou and his people—they never fell in. And I realized that at our organizational meeting. We had assembled a board now. I assembled a board and wrote a business plan. Got us in front of Georgia Tech's business incubator. They have an incubator called the Advanced Technology Development Center. We were trying to do something and we needed resources. Because they have a place where your company can go to, and that allows you to share resources and ideas and to get on the agenda with incubator. Everybody wants in, and so you get to go and pitch them. So I'm like, okay, I don't necessary agree with you, Lou, but this is your show.

Derrick did not think of himself as an entrepreneur or businessman. But he was certainly more of one than Lou Macalou and the other folks on his

UBP team. Derrick though that Lou's way of organizing not just UBP's work, but his management of his team, was all wrong. He seemed to fuck up everything from managing what little finances the team had, to assigning tasks and responsibility among his fellow engineers. Lou recognized as much. But he was the master coder, and from his point of view, UBP was his to own and showcase. But he still asked Derrick to pitch UBP to potential investors. Derrick agreed to do this, but not before making his concerns known to Lou.

So anyway, I pitched the material that we put together. But I said to Lou, "Let me tell you something, since you guys can't seem to get along, I understand that I'm not . . . that this thing is not going to happen." But it happened. We made the presentation. But the director of the Incubator says, "Look, guys, I don't think we can get behind you. Because this team is a little shaky. But I'll tell you what, if it doesn't work out, you call me, Derrick. Because we'd like to have you over here." So I said to myself, "Well, we're in the water now." I said, all right, all right, cool.

So now we're back in the organizational meeting in March of '96, and I didn't realize at the time, here's what happened. Lou—aerospace engineer, programmer extraordinaire—he's built all the code for the entire site of the Universal Black Pages. He's also written a very nice search engine that I would say rivals anything that Yahoo! had built. He wrote it back in 1994. It was very simple, very simple. So he suggests to Larry, who was an electrical engineer, like myself. But we're in different areas, I was in signal processing and Larry was in micro-electronics. So we are in the same major, but we're still on different ball fields. So Lou starts this conversation.

Larry—hey, I think we're going to use this type of web server.
I [Derrick] concede, Lou. Larry, that sounds like it will work.
HOW DARE YOU SUGGEST WHAT KIND OF WEB SERVER WE SHOULD USE!

The next day Larry walks into the office. I can tell by his face. . . . So I'm like, o-o-o-kay, we can put the champagne—over—there.

Why are you quitting?
Well, Lou is crossing over into my lane. He is not a server person. We see him breaking the bank. I'm out!
Larry, come on man. Wait!

Fortunately, Lou showed back up the next day. So we go in there and make a presentation to the national black data processing associates. The BDPA back in those days was, I guess, what you would call the older version of the Society of Black Engineers. They represented engineering and learning more on the old school data processing and computer needs. It was a great group, great group.

So we made a lot of inroads presenting there, and we were able to forge ahead, attract some more investments, which is the only way we can move forward in comfort. We had this elaborate stock agreement that Lou and Larry had put together—for a company that just raised its first capital. They've got this agreement that says they would both always have 47.62 percent of all stock. . . . So if any additional stock is authorized they would both be issued, we'd be issued amounts of stock enough to maintain their respective numbers and control.

And I'm like, "Who came up with this?" The eternal question. . . . Let me answer it for you—some engineers. This is crazy. So I go over to them, and I said now tell me what happened. They were at the same school. They didn't really know each other. They both come from MIT and here is the nature of the relationship. In essence there was no relationship. I go up to Larry and have the same conversation, and it's easy to see. Not only was this never gonna work. It's not gonna work now.

So I wrote to the investors, who happened to be my best friends, and I said, "Okay, guys, this is the story. . . ." And they said, "Well, look, we support whatever the project is. We just want you to know we gave that money to you." That you is me. "So you need to figure out how you're going to get our money back."

So I call Lou up.

Hey, Lou. I've got to tell you something.
    You better let me go first.
Damn.

Derrick is afraid of what he knows is coming.

    I'm in San Diego.
Why are you in San Diego?
    Well, I needed to take a job, so I'm here now. It's my first day.

And then he didn't tell me anything more. Come to find out, there's a lady involved.

Figure 8.3. Derrick Brown, inside the office of the Universal Black Pages/ Knowledge Base at Georgia Tech University. Courtesy of Derrick Brown.

So now it all makes sense. But everything that we had is now in my lap. And I said, "Well, in the greater scheme of things, I guess this is what I do."

So I learned. Every part of operations. How to program, that wasn't my thing. I continue that process of reviewing the sites, writing the descriptions and building a database, because that was our capital and that's also my sweet spot of interest. So now the website has almost five thousand links. I just had no idea that there's that much stuff out there about us. Because it's still not that easy to make a webpage in 1996. Not everybody knows how to do it. It's not something people have money to do. I'm interacting every day with people all over the world and I'm smiling on the inside because I'm like, okay, these guys are walking away from all of this.

But I guess the joke was on me. The whole thing got to a point where it was such an overwhelming experience. It's information overload. You've got so much stuff coming at you, and it's all exciting you, it's stimulating you. Your mind's moving in a thousand different directions and creates stress. I was stressed. I was stressed. So I said no. I might be veering too far in this direction, and getting away from knowledge-based directions.

Chapter Nine

ONE HUNDRED YEARS BLACK

A CAUTIONARY TALE

Yesterday. Today. Tomorrow. With these three words as its theme, the Congressional Black Caucus opened its 26th Annual Legislative Convention on September 11, 1996. Sony Music executive and Congressional Black Caucus Foundation chairman LaBaron Taylor opened the conference.

We are constantly being reminded of the parallels between what happened to America's "colored people" in 1896, 100 years ago at the end of the First Reconstruction Period, and what is happening to African Americans, as a people, in 1996.

No one knew this better than California congresswoman Maxine Waters. She had a story to tell. A story from black America's past. A story so explosive, so implausibly familiar that it cast a fresh cloud of complicity over the day, over the conference's agenda—over black America's technological future. The congresswoman's story wasn't about computers, or programming, or processing data.

It was about cocaine. And it was the perfect metaphor for black software. It answered that question, Who's going to pick winners and losers? Derrick or David? The rest of the Vanguard? Black folks? Or, someone else entirely? Congresswoman Waters needed to remind those who showed up to the conference to talk about the Internet that black software wasn't just about machines and cables, chips and modems. It was about America's fundamental, deep-seated and persistent first principles: white supremacy.

* * *

In the 1980s, Silicon Valley heralded the second high-tech revolution. From its Bay Area core, the region radiates outward, from Stanford University to the west, and up to Daly City, down past San Jose, and over to Hayward. It was named for invention, innovation, and entrepreneurship. The region had long ago played midwife to industry. It helped to birth the microchip, ARPANET, disk drives, and the personal computer. But in the 1980s cocaine was the valley's newest, purest, most preferred, and best-distributed high-tech curio.

You see, the valley sold dreams. Its sprawling intellectual and industrial spaces, including its Stanford-connected labs and government-sponsored research centers, provided a new frontier for imagination to wander. Each fed the impulse to build new tools with which to master the universe. Its financiers capitalized investment in its fantasies. The new tools brought to market in droves provided the satisfaction that comes from dreams not deferred. Cocaine was tailor-made to fit the valley's technological and entrepreneurial ethos, its daily grind, its demand to create value, its pervasive drive to succeed, and its capacity to aspire.[1]

Paradise White came to the valley produced from raw materials. The main ingredient included the alkaline substance found in coca leaves. Some people chewed the leaves to extract its intoxicating ingredient, but powder cocaine had to be manufactured. After harvesting, you produced a mash from the coca leaves by adding water, lime juice, and diesel fuel. You filtered it, and a dark, thick paste called base remained. It resembled a piece of decomposing feces plucked from a pool of sewage the day after a heavy rain. To purify the base you added sulfuric acid, which turned the cocaine base to cocaine sulfate. You mixed in, then filtered out, ammonia or potassium. This produced a yellowish-brown solution that had to dry. Then, you dissolved the paste into ether, mixed it with hydrochloric acid and acetone, and filtered and dried it one last time. By then you had finally transformed that dark, murky paste into an immaculate white powder.

But cocaine is a chameleon. Its identity, character, and reputation reflect those who have access to it, who can afford to use it, and have cultivated a need to exploit what it affords. It courts those with power and resources. It beckons them to alter its chemistry, enjoy its leisure, shape its expressive environment, and determine whose interests will be served.

In the early 1980s, one gram of Paradise White sold for about $625. Silicon Valley was the color of pearl, and its demographics were destiny. Except for Alameda County, white people dominated the valley. Each year, its residents raked in more than $100 billion from wages and added income from dividends, interest, and net rental income to their cash stockpiles. These were educated men and women. They lived in neighborhoods where almost every home was both occupied and owned. Those homes cost an average of $250,000 a pop, yet one out of every five homeowners didn't even need a mortgage. Most of them made things, sold things, managed those who made them, fixed the things they made after they were sold, or provided professional services so that all the things worked according to their design.

The valley sold dreams. And in the early 1980s, its engineers and scientists, venture capitalists and salesmen charged headlong into a competitive circadian rhythm. They bathed themselves in this pristine white powder like it was manna rained down from heaven after forty years of famine. Cocaine smoothed rough edges formed in what became a ruthless, high-tech hustle to out-think, out-research, out-invent, out-develop, out-produce, out-market, out-sell, out-distribute, and outshine their competitors. They were risk-takers, the kind even professional actuaries didn't want to bet on. *If you do a lot of business in the executive professional circles, particularly on the two coasts, and doubly so if you deal with "high tech" engineers in Silicon Valley . . . you doubtlessly have a fair number of cocaine users in your standard risk subset!*[2] a conference speaker once pointed out to a room full of insurance professionals.

Cocaine represented a look, a lifestyle, an aesthetic that revealed naked class affinities and aspirations. Take for example the "Blue Lady." Back before the federal government criminalized drug paraphernalia and advertising in the early 1980s, the Blue Lady filled a full-page magazine spread. Her eyes; the negligee that barely clung to her breasts; the tiny, nickel-plated handgun she clutched in her hand—all invited the browser to partake of what lay between her legs: a pregnant, plastic baggie filled with powder alighted atop a silver scale. The Blue Lady hawked Italian *Mannite Conoscenti-Blue*, a mild baby laxative used to cut cocaine.[3]

Cocaine sold itself. But it needed an image to maintain its price and solidify its place as the drug of choice for those with exquisite taste. *When you have worked very hard and succeeded, you sometimes enjoy buying something simply for its beauty. It's for those individuals who appreciate beauty for beauty's sake.* The ad for a product called *Ivory Snow* explained as much. A company called L. Bandel, out of Los Angeles, California, sold it via mail order. *Each of our exotic spoons, straws and vials are delicately carved, by skilled artisans, from the finest center cuts of imported African Ivory.*

One dealer from Buffalo, New York, even sold a set of 22k gold-plated razor blades. He advertised it as the *gold standard*, in a headline sandwiched between two lines of coke. The copy described the *unmistakable elegance of gold blades designed for the discriminating player.* The mirrors used to cut cocaine were branded as *White Lady's.* Stone surfaces used for the same purposes were called *polished.* To maintain *purity*, one seller from Southgate, California, even sold the high-tech *Hot Box. For less than the price of two grams, you may never buy a bad gram again* the caption read. The product described itself as a *scientific instrument that brings the expensive laboratory into the home and allows you to identify, describe and separate the percentages of common adulterants used as cuts for certain crystalline materials.*

New York City's high-class nightclub scene stoked cocaine's image as a drug of the Hollywood celebrity set. That image was already in place by the time it became the drug of choice for Silicon Valley's high rollers. Popular lore abounded with incredible tales. The valley's would-be titans of industry preferred their cocaine at the office, or at house parties where husbands gathered together to talk incessantly about computers, while ignoring their wives. The valley had developed an image that it was the kind of place where even the children of Silicon Valley executives would *watch their parents hit two lines of cocaine before trying to swing a deal.*[4]

Cocaine retained every bit of its glitz and glamour. But in the valley, it was all designed to push the work. Cocaine labored in service of the dream. For most, the dream was a fantasy, but they chased it nonetheless. Cocaine kept them in the race.

You look around and you see your neighbor who has made millions in a public offering and you feel he isn't any smarter or more technically knowledgeable than you are. And it seems absurdly easy to do what he did, an industry executive explained at the time. *But if you don't do it now, you might not have the opportunity ten years from now. You cannot relax because every Monday morning in the* San Jose Mercury *there's another story about somebody who made it that hits you right as you're drinking your coffee. The only way to do your job is to work an eighty-hour week.*[5]

The man who conveyed this story represented the Silicon Valley rat race realities that led to cocaine's explosion. Drug sales and use ran rampant at work. Everywhere from IBM to Syntex, Lockheed to Santa Clara County itself, discovered that they had problems.

The US Drug Enforcement Agency (DEA) and FBI were firmly entrenched in Silicon Valley. Year after year they seized cocaine shipments. Stashes ranged from ten to hundreds of pounds. Traffickers and high-rolling dealers were sometimes arrested; sometimes they were jailed. Every now and then one was even prosecuted. But everyday cocaine users—even when their employers or the law were fully aware—found safe haven. Boutique hospital drug rehab centers nursed their disease, one that therapists said was out of the addicts' control. When it wasn't a glaring problem that exposed them to liability, workplaces turned a blind eye.

Cocaine was a white drug, designed for white people with the privilege to dream, and the opportunity to pursue those dreams.

* * *

Things were different 350 miles to the south. In the 1980s, Silicon Valley had climbed steadily toward its technological zenith. Meanwhile, South Central Los Angeles rushed toward a kind of hell from which no one could soon or simply resurrect it. It was black, intensely segregated, and overwhelmingly poor. Educational and economic opportunities were pipe dreams. Since the 1920s, government officials, real estate magnates, city planners, bankers, and other powers designed South Central to be precisely what it was. It was built to contain an urban wasteland. The powers that be did not want its residents to infect the rest of the city, or the country for that matter.

Cocaine had spawned a lifestyle, a culture, axes of racial and class associations, markets, economies, institutional networks, and a host of unintended consequences. Crack is derived from cocaine. To create it you dissolve street cocaine in water, add baking soda, and cook it down to its chemical base. You let it dry, then chip it into pebble-sized rocks. That simple and cheap process yielded a product that was more than twice as pure as street cocaine. It delivered a quicker and more intense high than cocaine, but it sold for as little as $2.50 a hit.

Cocaine's complexion changed dramatically with its technological transformation. Sure, crack found a home in white communities across the country. In fact, the DEA claims that when crack first arrived in New York City, middle- and upper-class whites flocked to the drug first. But crack's widespread accessibility—compared to cocaine—created vast, new markets, primarily among black and brown people in central cities across the United States.

Crack was a black drug. It was designed to satiate and devastate communities full of people whose dreams always seemed not just to be deferred, but crushed altogether.

* * *

Code is the basic building block of computer software. It is a language that binary computing systems understand. Software itself is merely a set of instructions that a programmer inputs into a computer. The machine processes the information, performs the instructed task, and produces a desired, pre-programmed outcome.

Today we typically associate code with computing and information processing. But code is the foundation for many other systems. Computer code tells machines how to work. Genetic code tells our skin what to look like. Legal codes instruct us about how we can and cannot behave.

In some cases people design code—software—to create an array of potential user applications. Microsoft designed word processing software, for example,enabling a user to formulate a letter, design a graphic, build a book index, format a book, or even write software. This is the kind of code we might associate with cocaine. Cocaine was designed to provide pleasure, stimulate entertainment, prolong mental acuity, and allow people to stay awake longer and work more hours in a day. Its comparatively limited policing— by legal codes and law enforcement, workplace policy, or social convention—provided a sense of freedom and opportunity. Silicon Valley associated cocaine with professional and economic success.

In other cases, people design software to target and solve a specific problem. Antivirus software, for instance, is designed to identify malignant code on a computer. It isolates it from the system, removes it, and then prevents future infections. This was the crack code, used to identify, isolate, and remove America's greatest problem: black people. The code in which it was written was legal, tactical, and racial.

Cocaine fashioned a lifestyle for the white, rich and famous. As long as it remained so, law enforcement remained relatively carefree about its widespread use. But crack was practically born black. Black infants sometimes enter the world with much lighter skin that darkens with the day. Similarly, the technology transfer between Bahamian cocaine dealers and US traffickers found its first home in Miami, Florida.

People who had been to the Bahamas, or Bahamians, then migrated to Miami or Los Angeles, carrying this drug know-how [how to turn powder cocaine into crack] *with them. When Miami police first busted a "rock house" in 1982 the scene was still reasonably respectable.* People of all races, classes and genders partook in the drug's pleasures. *Soon rock, or crack had spread to the traditional drug markets of Miami's poor, Black neighborhoods . . . where local youths learned how to cook up crack from Caribbean immigrants and went into business.*[6]

A *New York Times* headline announced in 1985 "A New, Purified Form of Cocaine Causes Alarm as Abuse Increases."[7] By then, crack was already seen as an inner-city (poor, black) drug. And two things became crystal clear. Crack bred violence and crime. In the popular imagination, cocaine users were addicts. They were white, sick, and suffering. Crack users were sick, too. When crack first hit the streets, pharmacologists, doctors, and public health officials practically shouted from the rooftops, telling the world what would happen if freebase cocaine was transformed for mass distribution.[8] They claimed that crack's intensity and addiction speed, compared to cocaine, produced an outsized need to obtain the drug. Crack use and violence began to skyrocket in 1985, with already-poor users turning to robbery in order to obtain the drug. This was just one year after the Comprehensive Crime Control Act of 1984 (3CA) set the stage for the new drug war and one year before the Anti Drug Abuse Act of 1986 (ADAA) initiated law enforcement's critical escalation.

Crack, blackness, and criminality aligned. They formed a composite image in the American mind. The federal government designed and developed a very deliberate, targeted, and ruthless crack strategy based on that image.

The code enabled law enforcement to profile and target young, black men for arrest and incarceration. It was efficient, elegant, legible, flexible—all characteristics of great software programs.

* * *

"You stand accused!" Congresswoman Maxine Waters shouted into a dozen microphones huddled together at a ballroom podium. Her words exploded with contempt. Her eyes saturated with rage. Her pointed, three-finger jabs darted outward with disdain. She directed them all like fiery bolts of lightning, hurled toward shadowy figures not present among those who filled the conference room on September 13, 1996. She aimed to start a political brawl—among the Congressional Black Caucus, her South Central constituents against the CIA, law enforcement, and every public official who carried out orders in Ronald Reagan's drug war.

For the preceding month, the *San Jose Mercury News* had rained hydrogen bomb–like headlines on South Central Los Angeles, and the whole of black America. They connected all the dots, stating *"Crack" Plague's Roots Are in Nicaraguan Drug War; Colombia-Bay Area Drug Pipeline Helped Finance CIA-backed Contras; 80's Effort to Assist Guerillas Left Legacy of Drugs, Gangs in Black L.A.*[9] A large photograph placed just right of center featured the key connection between the Silicon Valley supply chain, and its South Central destination. They called him "Freeway" Rick Ross. The page featured his two accomplices as well, Danilo Blandon and Norman Menesses.

Menesses was a known Nicaraguan drug trafficker. He oversaw massive cocaine shipments from South America into the San Francisco Bay. Blandon was a Nicaraguan freedom fighter desperate to raise money to fund his cause. Thus he brokered the sale of those massive quantities of cocaine to Ross.

Ross hailed from Watts, in South Central Los Angeles. Once a college-bound tennis player, Ross thirsted for wealth. He also had a natural entrepreneurial gift. Ross bought the cocaine from Blandon on consignment and he cooked it in any number of rock houses he'd established in South Central. Then he distributed it, and made LA's street gangs critical links in the chain. Ross made millions, while Blandon shipped his proceeds back to Nicaragua, and the process repeated. This was *the Reagan Doctrine*, a presidential administration that provided overt and covert assistance to guerilla movements throughout Latin America. Reagan aimed to quell what he believed was the nation's greatest threat: Soviet communism.

The story of the *San Jose Mercury News* special report that Congresswoman Waters seized upon continued for days. The paper created a special website to showcase the vast conspiracy. They titled the website *Dark Alliance*. The stories featured at *Dark Alliance* spread throughout the nation and the newly developing Web. And, it single-handedly raised the Web's profile among black Americans. After the story exploded, Malcolm at NetNoir acknowledged, *It's definitely affected the community. It's led to an awareness of the power of this medium. It's only part of a much longer process, but this is definitely a very important step.* Farai conjectured, *It's always a chicken-and-egg situation. Are black people not on the Web because there's no content for them, or is there no content for them because they're not there?*

If any content was going to attract black people to the Web, the story that Maxine Waters and Dark Alliance told would be it. One could draw one of two conclusions from the *Mercury News*'s report. At worst, the US federal government flooded South Central Los Angeles, and every major black community in every major city across the country, with crack cocaine. At best, the federal government negligently stood silent while crack unwittingly wreaked havoc on black America.

Congresswoman Waters rallied black America, her colleagues in the Congressional Black Caucus, other congressional and Senate allies in Washington, DC, and the legions gathered to protest back in her home district. She forced hearings. She forced the CIA to account for their deeds. Waters and her allies felt at least a little vindicated. If nothing else, it showcased the US government's continued willingness to deploy any and all technology at its disposal to maintain America's prevailing racial order: to keep black America in its place, and in its sights.

* * *

The way Congresswoman Waters opened the 1996 conference didn't bode well for Derrick. He had already begun to feel hints of strain. He had reluctantly taken the reigns of a company that forced him to seek profit at any cost. He had investors to pay back. BGS Infosystems—that's how Derrick ended up in Washington, DC, in September 1996 for the Congressional Black Caucus Foundation's Annual Legislative Conference. It wasn't quite the gathering of eagles that had descended on Washington two years prior. That event had channeled a full day's thinking toward how to build onramps to the information superhighway for black people.

All the players in the new black technology elite were there. David had made millions by this point, and NetNoir was still building itself. GoAfro limped along. People at least still spoke of it when they listed the top three destinations for black people on the Web. And Larry Irving was there.

From the get-go, Derrick got lost somewhere in between the two tech kingpins' agendas. David gripped the levers of the dotcom boom, while Irving's unseen hand shaped the federal government's tech policy agenda. Derrick arrived after Congresswoman Waters's explosive announcement on September 13.

It was an absolute mess, and I have not seen David since then. Tupac died that weekend. We were driving over to the meeting, we find out that he had died. But it was just a mess, man. Everybody in that room was trying to be the don gotta or somethin'.

But my favorite collaborators were Rodney Jordan, who had started Melanet. com. And Anita Brown. And you won't get to talk to Anita. But she had so much more to do with this than any of us understood, because she was not a techie. She was the leader of Black Geeks Online. Black Geeks Online was an entirely email-based organization, a listserve that was run by Anita Brown. She was no relation to me, but we still got really tight, because we were both Browns and surfed the Internet twenty-four hours a day. And she was cool with me. She was cool with David. She was cool to Rodney. She tried to get us all together all the time. And she had this humungous email database and we were like, "How in the world could you collect that many email addresses?"

And it's like, she said she had 25,000 emails, and she would email them. I met so many people just reading her emails. So yeah, she was the connector. She had so much more than she had even told me. She just dropped name after name and story after story. She told me, "You guys are really in a position to, like, run with this stuff, because you know what it's all about at its deepest and most intricate level."

Anyway, back to the main subject—the Congressional Black Caucus has an annual conference and we met at that conference, I forget who sponsored it. But it was just a meeting to outline the black Internet agenda. And it was a very self-important and heavy kind of atmosphere. So from the jump, I wasn't feeling it. I really wasn't feeling it. So that probably colored everything I heard that day. But much of what transpired was we organized committees and we met all of the players, many of whom we hadn't seen face-to-face yet. And everybody carried themselves like it all rested on their shoulders.

But really, it didn't rest on anybody's shoulders, and we're all trying to figure this out. This is such a new thing. None of us really know what to do with it yet. So the people you end up creating the greatest cache with are the folks who are honest enough to say, "Man, I don't know what to do with this stuff yet. But the possibilities are ginormous. Let's get together and see what we can figure out."

So for us, that trip was probably a disaster. That's a safe word. It's not the softest word. It's a safe word.

Derrick spoke for himself. And those he determined were his people—people motivated by black community uplift. That did not include David.

But David had always been very clear and unapologetic about his motives. *I believed that the best way to do for my community was to do for myself.*

Their two positions were not necessarily at odds. But if you were Derrick, it was certainly easy to see things that way. Especially because the escalating tensions about BGS Infosystems quickly drove Derrick to a crossroads.

* * *

Things began to take a turn for the worse after the 1996 CBCF gathering in Washington. And not just for Derrick; the industry, which he helped to create, was changing. David and Malcolm were pivoting. It was starting to feel like black America's revolutionary movement online was becoming just a very short moment in our history.

NetNoir and GoAfro locked in their bicoastal tussle. Just two years out from launching NetNoir, both Malcolm and David had already turned their eyes toward what was coming next.

My job was to build the brand. That's when I got a lot of personal attention, because I was involved on every talk show, every cover magazine. At the same time, Larry Irving, who was assistant secretary of commerce under Ron Brown before Ron was killed—Larry was pushing the digital divide and that old phrase. And that's where it may help to exploit and promote, and that's where things got pushed to the medium, especially in the early stages of the Internet. The haves and have-nots, the digital divide, who will have access to all this information in the future and technology and who will not?

And so then, the next sentence of the next paragraph would be, "But there's this company called NetNoir that's providing content." So we got caught up in that kind of dialogue, and it was strange for us a bit because we were just doing our thing. Let's say we leveraged it. Because all press is good press, right? So I'm trying to build my media brand called NetNoir at the time.

So we were able to leverage the digital divide discussion, but I pulled away from it because I was like, "Well, wait a minute. We also have cross over here. This is just showing that our culture is here and it's vibrant and will survive and thrive in this new digital medium." You guys are trying to figure out how people of color will get access to the Internet in the South Bronx or rural whites will get it in the middle of Iowa. Well, that's an interesting discussion. It's a very good policy discussion, but I'm a media company and I'm trying to make money and be a successful capitalist.

By 1999, even Ted had moved on. Walled garden portals like AOL could no longer hold back black masses gathering on the open Web. Malcolm set his eyes east to Asia. And David was about to call it quits. But before everything began to fade away, another player began edging its way into the black Internet market.

* * *

Bo's a Rarity; Few Pros Reach Two-Sport Stardom, read the *Seattle Times's* sports page in 1988.[10] *Shaq Just Plays It Cool in Orlando*, topped the *Chicago Tribune's* sports section in 1992.[11] The same paper in 1993 flashed *Looking for a New Computer? It's in the Mail; Ordering PCs through Catalog Is Popular, but Drawbacks Are Emerging* across its North sports section.[12]

The headlines reflect the abrupt shift that computers made to every industry, especially and including the newspaper business. They also say a lot about the writer's shifting interests and the millions of people who read Barry S. Cooper's words.

Barry worked as a sportswriter for the *Orlando Sentinel* in 1994. But he realized that connected computers were about to change the newspaper business. He also realized that he just might take advantage of the new digital tools, and that they could bequeath to others some of the same privilege that he enjoyed. This was the privilege of having a voice.

I was a sportswriter for a number of years, and wound up migrating to AOL by the hour. And African Americans overindexed in their use of AOL. It was largely chatrooms and news, and folks were paying 3.99 an hour, and racking up huge bills—four to five hundred dollars—because it was so addictive. At the newspaper, I was working putting the newspaper online, and because of the activity on AOL I saw the opportunity to create a community online that would be for and about African Americans. That's why I moved away from traditional journalism into a more community-focused effort.

I did it on no budget, just a shoestring, using volunteers who worked for no pay. These were African American men and women who liked the concept, liked the idea. They were thrilled about having their voices heard and building a community. So, I did my day job at the newspaper, putting the Orlando Sentinel *online, and at night with my volunteers we built it.*

We were lucky because the technology then was not that sophisticated at all. We used AOL's proprietary platform that was called Rainman. And so we used their platform to build our chatrooms, post our content. It was relatively easy. You didn't need a technical background to work on it. And so we used those tools to get the site up and later we moved to the open Internet and had a website. But the barrier to entry with regard to technology was very low back then.

Barry and his professionally unpaid team of new, black, community journalists had realized their collective vision. They had started to build their own little neighborhood within AOL. Suddenly they felt like real estate agents looking for clever ways to draw potential buyers to new neighborhoods. Barry and his team needed a name.

We loved the name from the start. We thought it was in-your-face. It was dramatic. It was clear. And once I coined the phrase, we went with it and we didn't look back. A lot of it too was that we were gonna be urban, and we were gonna be for everybody who was African American or who appreciated the African American spirit and lifestyle. I wanted a short name. And that's why we settled on "Black Voices."

I really stumbled into this, just because I loved the community. And I was really a fan of the AOL platform and the emerging technology, and I wanted to play a role in it. And I never thought about making money. It was always about the love of being involved in it.

That moment was short lived. Black Voices was up and running. People—not just users, but people at the *Tribune*—began to notice. The *Orlando Sentinel* had included a button link to Black Voices on its newspaper's AOL site. The corporate bean counters saw what was happening right in front of their eyes. The link wasn't steering traffic from the *Sentinel* to Black Voices. Black Voices flooded the *Sentinel* with new black visitors. The new feature boosted overall usage of the *Sentinel*'s site from 60,000 log-on hours a month to 100,000.

Soon, Black Voices stumbled out of AOL's provincial walls and out onto the open Web. The new world wide web site featured much more interactive content. When it first hit the Web early in 1998 Black Voices' front page featured links to relevant news: news about a black newspaper in Mississippi that had been firebombed by white supremacists; news about Los Angeles Lakers star Kareem Abdul-Jabbar settling a civil lawsuit; news about Hillary Clinton choosing to *stand by her man*. But it also solicited users' opinion on popular questions.

Your thoughts please, on black men and white women?

The site drew people toward its chat services.

Afro Chat Is All the Way Live. Come on in!

The front page highlighted a "member of the day." On January 28, 1998, that was a gentleman named A. Byrd—who used his featured spot to tell the world about himself.

Just a goodtime Adventurous Guy, looking for spontaneous woman, any/race to communicate with, get serious about, then get romantic with, somewhere in

Paradise. Bless all the Beautiful women in the WORLD. . . . AND THAT MEANS YOU.

The site featured the best of the online experience as it existed up to that point. That included Black Voices' *Black Wall*. The wall featured message boards, curated by topic.

Living and Loving. Black Sexuality. Ebony Gay & Lesbian. Roots. Teens Represent. Ticked Off. Money Talks. Going Places.

Young people could even go to the site to swap pictures from Freaknik, the legendary black spring break party held every year in Atlanta.

One of our biggest successes was in utilizing user-generated content. I remember us doing member photos, where you submitted your photo, and the photo would appear in the Black Voices site. And every day we would highlight a member of the day. We were very early in a number of ways. Once we moved the operation to Chicago, we would send people out with video recorders, and stop African Americans on the street and ask them their opinion about something that was in the news that day. We would use blog content from people who would send in essays and stories, and we published that as well.

Barry and Black Voices moved to Chicago because the *Tribune* decided that the site had created a market worth investing in.

It wasn't until 1998/1999 when we approached the Tribune *about funding this, so that we could have a real staff. And they invested $5 million. That's when I started thinking about making money. That's when the focus was put on running a profitable business. So yes, I started out just because it was a great hobby. I thought it was a neat idea. Let's have some fun with it.*

We were nestled inside a conservative media company that had never really targeted the African American community. I think the Tribune Company had some investment in Soul Train, *but that's as deep as they got. So we were really new to them as a black product. But mostly we were left alone on the content side. Largely because we were having some early success.*

By the end of the 1990s Black Voices entertained nearly a million regular visitors. Its popularity lead General Motors to announce a multimillion-dollar marketing deal with Blackvoices.com in 2000. The deal delivered Black Voices' consumer market directly to the automaker.

The online world at the turn of the twenty-first century was big enough for two larger-than-life sites targeting African Americans—two sites that featured the same basic content and features and two sites located within what was left of AOL's portal.

The commercial prospects for folks like David and Barry, Lee and Malcolm, Timothy Jenkins, and other black Web entrepreneurs began to shift. NetNoir swallowed up GoAfro, even before AOL acquired CompuServe in 1998. Black Voices thrived. But the *Tribune* could not figure out how to actually make it profitable. Some in the market said the dotcom boom was one giant bubble just waiting to burst.

Still, the Vanguard remained. Though, anyone really looking could see they were starting to fade. Separately and together, they had built something. But they may have left more unanswered questions a decade after it all started, and more than they encountered when they had begun.

Back in 1994 at the CBCF conference, everyone raved about the information superhighway. Less than a decade later, its first on-ramps, interchanges, and meaningful exit points for black folks started to crack and crumble. And, the related, but larger question still loomed. This was the question that Congresswoman Waters prompted when she reminded them that the nation's enduring legacy of slavery and white supremacy shaped everything about black life, from the drug war to the tech wars. The Vanguard had to grapple with the question, What would and should black America's relationship to technology, to computers, to the Internet be as we approach the beginning of a new century?

Figure 9.1. Barry Cooper, founder of "Black Voices." Courtesy of Barry Cooper.

Chapter Ten

TAKING "IT" TO THE STREETS

By 1998, Derrick had finally found his footing. It had been no easy road, and he had struggled to lead and build the UBP single-handedly. But focusing on Knowledge Base led him back to his true identity, an identity more consistent with his name, his conscience. Now his path forward became clear, and he made it known to all those whose online path led them across UBP.com.

November 27, 1998

To Our Loyal Supporters:

Since 1994, The Universal Black Pages (UBP) has sought to present the most comprehensive catalog of the African Diaspora's Internet/World Wide Web offerings to our audience. As many of you have undoubtedly noticed, the quality and timeliness of our publication has slipped greatly over the past two years. Many administrative and technical shortcomings led to our being suffocated by our own success.

Our once-mighty editorial team has been reduced to a staff of one (me), and since I am now actively involved in the day-to-day operations of my nonprofit organization, KnowledgeBase, Inc., I have been forced to streamline The UBP's operations so that we can once again produce a quality publication. The first step in that operation is to completely edit and update the current version of The UBP (which will take quite some time), after which I will begin to edit and add the nearly 50,000 links we have received over the past year.

You may ask—what can I do to help? Well, you can help by continuing to visit us regularly, and indulging us with patience as we handle our business. As you use the page, you may encounter dead links, or errors of other types—if you do, *please let us know*. We want to provide a quality information service to you, and your continued support can help us to reach our goals.

Once again, thank you for your patience, understanding, and—most of all—your past, present, and future support.

Sincerely,

Derrick Brown

Editor, *The UBP*

Executive Director, KnowledgeBase, Inc.

Even as Derrick began to let go, the Vanguard had been summoned, yet again, to Washington, DC.

This time, Bill Clinton's White House called. They called it a briefing. There was little room for discussion, debate, and making plans. Those days ended with the CBCF meeting in '96. Regardless, this was the moment when Derrick believed he had—finally—found his people. There was no drama. No one was killed. There were no conspiratorial bombshells. No one jockeying for position. It was just a time for Derrick to take it all in, converse with his friends, take stock of where he was, and reflect on where we stood as a people. He contemplated what our relationship was, what it could and should be to the Internet, and to computing technology in general.

And, after the meeting was over, Derrick received an invitation to publicly reflect. He offered what he saw as our way forward.

10 September 1999

Andrew Glass

Senior Correspondent

Cox Newspapers Washington Bureau

400 North Capitol Street NW, Suite 750

Washington, DC 20001-1536

202.887.8318, 202.331.1055 (FAX)

andyg@coxnews.com

Dear Andrew,

I just wanted to share some thoughts with you about the two-day briefing after I had taken time to pause and reflect...

The White House briefing was organized in the midst of growing public concern over the Digital Divide. The group invited to the briefing represents the vanguard that has already been working for several years to conquer this divide by creating compelling Internet content and by organizing substantive infrastructure-building and training efforts targeted at African-Americans.

My experience here has convinced me of certain truths that are evident and critical to the long-term impact our leadership can have in working to reduce the Digital Divide. Our efforts have to view computers and information technology as *means* to produce the fruitful *end* of productive, skilled, purposeful people. To produce this end, we have to prioritize our investments:

1. Our initial investment must be to foster a stronger sense of community and collaboration amongst the leaders of this movement—i.e., we must all see the same end, and channel our collective efforts towards that end.

2. Our next investment must focus on building infrastructure (creating technology centers in churches, schools, and community centers) that grants access to disparate populations, and training that makes this access meaningful.

3. Further investment must be made to open access to capital markets that will allow our leadership to build the capacity necessary to make significant impact. We must teach each other the leadership and entrepreneurial skills necessary to gain access to these markets.

4. All subsequent investments must be made to continue developing compelling content targeted at these disparate groups. This content must be produced by companies that we collaborate to build.

Feel free to contact me with any further questions regarding the briefing via return email.

Sincerely,

Derrick Brown

Executive Director, KnowledgeBase (http://www.ubp.com/KB)

In 1999, Bill Clinton, preceding his last year in office, organized a similar event of the African-American Internet vanguard. And who's to say? I'm twenty-nine when this rolls around; I was twenty-six for the first event. Might have been a little wiser second time around.

That event was one of the greatest things I've ever gone to in my life. Because we were there for two days and I actually wrote a press release that the Atlanta Journal-Constitution *ran verbatim, because they were not allowed access to the event. And I think on the last day one of their reporters got in there somehow but he didn't know what was going on. So I gave him his story. And I'm a writer. So it's like, yeah, I'm putting together my notes because there's an agenda that I've created here that I want to disseminate, and just like you need a story, I need some ink. So there's your story.*

And he ran that. That did a lot of things for me, because he ran it verbatim. Verbatim. And if you let me write my story, I'm going to come out looking pretty good. And I'm going to say what it is I'm trying to say. And I articulated I think a four-point agenda for where we go from here. And the first part of that agenda was what Lee Bailey told me back in 1996. Content is king. You have to gear all of our efforts towards engaging, compelling content, because at the end of the day this is kind of like a class, getting people to learn about this stuff and learning how to leverage it.

It's the class that you're required to take and because you're required to take it, you're not that interested in it. So you're going to have to figure out how to entertain folks with this stuff. Not entertain, not education. You have to do both of them at the same time where if you're not into education, you're not totally there for entertainment either. You've got to, you've got to figure out how to do this in a creative way. That's first and foremost, and that's probably still the first bullet in the agenda.

Back then the second bullet in the agenda was access. Folks have to have computers, folks have to be wired. That's a big policy issue in 1999. I worked that with the Urban League a little bit, that next year. Democratizing access. That's not as big a deal now. Still big deal, just not as big a deal as it was fifteen years ago. And third point was collaboration. As bad as it is, we worked together in 1996 as badly as the Universal Black Pages Team fell apart. We've got to figure out how to make some synergy out of this, because there's a wealth of talent in our community.

It's a small group, but it's a talented group, and we figure out how to bring that group together, we're going to be able to do something. And then the last point was capital. Capital. Because I was taught, at Business Incubator, you take care of those first three things, somebody or some entity is going to find financing for it. By the time you do all that legwork on those first three points of the agenda, you might trip over the money. I saw lots and lots of money at that White House briefing.

I saw Courtland Cox, man, I've met Courtland Cox at that briefing. He was the head of the Minority Business Development Agency, which I think is part of SBA. So this is a real, live civil rights icon who has engaged in the struggle. He's been part of SNCC. And now he's in the Clinton administration. That was like, hey, wow, oh, whoa. So that White House trip, man, that was the bee's knees.

So at this point, man, even in 1999, I said, well, life has been no crystal stair. But if I look at where I've come from, even since 1994, part of the Black Graduate Student Association, it's like, hey, I'm not rich from any of this, but I do have a wealth of knowledge and can see so far down the road now that there's just so much more to all of this. And this concept I have around KnowledgeBase paid the same way. I get down the way I get down now because of what I learned. Not so much what I know, it's what I learned. Because there are far more things that I don't know than I do know. And as long as I'm out there learning, I'm amassing a level of knowledge that will allow me to take advantage of lots of things that I might see that other folks don't see.

For Derrick, the work continued. And he connected with new friends, and old ones—like Anita Brown—to keep it going.

* * *

Anita was old enough to be Derrick's mother. She didn't have a tenth of his education, and even less of his engineering expertise. But she had a gift for bringing people together, a desire to evangelize her own community to get online. And, she had the hottest email list in town. And by town, I mean the country.

WIRED magazine profiled Anita in 2000, inviting the Internet world to *MEET ANITA BROWN: The Best-Known Black Woman on the Web.*[1] However, for most of the century she helped to usher in, Anita spoke through her website: BlackGeeksOnline. That's where she told her story. That's where she shared with the world the reason that black geeks existed.

Black Geeks Online was launched in January 1996 to connect tech-savvy African Americans who are willing and eager to bridge the widening gap between technology haves and have-nots. As of November 1999 over 25,000 people have become members.

Why? Our experience indicates that from South Central to South Jersey computing is a hard sell in the 'hood. Unlike baggy pants, hip-hop music, and drugs, Information Technology (IT) is rarely marketed to African Americans. Black geeks rarely appear in media ads; there are few (if any) hardware and software ads in Emerge, Essence, Vibe, The Source, Black Enterprise; *and the nerd and geek images associated with computer professionals are still considered "uncool."*

Our "Taking IT to the Streets" offline expos have demonstrated that Net-literate volunteers make excellent trainers and role models: kids instruct other kids; women encourage reluctant women; and Black entepreneurs own the companies that donate the computers, multimedia software, dial-up service, etc. We urge volunteers to share their personal journey from newbie to netizen, explaining where they started and how they gained training, experience, and a rewarding career.

It's our aim to foster technology awareness, creating the aha! experience. We find out what a child or adult is passionate about and show them how the Internet can bring sounds, images, and information about Michael Jordan, Lauryn Hill, Will Smith, or Oprah right to the desktop. We use a culturally sensitive approach to introduce curious and often technophobic people to email, point-and-click, and surfing the Web. Then, because knowledge is power, we research and publish a "Taking IT to the Streets Net Guide" that points people to free email, low-cost computers, and to Net-connected libraries, schools, shelters, and other community technology centers where free and low-cost classes are offered in their locale. Once kids and adults get fired up about computers and the Net, the need increases for volunteers—virtual mentors, hands-on trainers, role models who can speak at churches, Boys & Girls Clubs, community tech centers, etc.

Perhaps Anita—affectionately known by her online and offline friends, family, and fans alike as *Miss D.C.*—had an inkling that she would one day be silenced, that her voice would no longer be audible, that her words would be hidden deep in archives that few people bother to dig through. Perhaps that's why she told her story, the blackgeeksonline story, the story of how she started a movement to *Take IT to the Streets.* She told that story so many times, in different ways every time. Each reflected a different and deeper perspective.

In January 1996, BLACK GEEKS ONLINE was launched by Anita Brown with eighteen men and women from the mid-Atlantic states who are committed to increasing computer literacy and Internet access among African Americans. Black Geeks Online is an Internet-based community organization that connects people of color from around the world. Our purpose is to share our talents and time to promote computer literacy and educate others about the power and potential of Internet technology.

We are by no means all *geeks*. Our members include educators, students, community leaders, entrepreneurs, and government officials. Our most active members are, however, new media developers, content providers, webmasters and others who are serious geeks. Charter members include Melanet, Black On Black Communications, NetNoir Online, Information Brokers, Inc., interCHANGE, AutoNetwork, and Keep It Real Online.

Mainly, we are connected digitally, sharing information and resources (new media developments, job announcements) and opportunities to share the technology within our communities. Our D.C. "chapter" has been hosting popular offline gatherings since August 1996. We're determined to get to know each other by more than just email addresses! Our members stretch across the U.S. and beyond to Hawaii, Alaska, Canada, Okinawa, U.K., Abidjan, and Johannesburg. Our newsletter, <bLINKS>, is widely distributed via email among African Americans, many of whom only have access to email—not the world-wide web. And Heads^UP bulletins announce job opportunities, scholarship and fellowships, conference and more. We offer commercial ads in these publications for a nominal fee.

Black Geeks Online publications are forwarded and reforwarded around the 'Net (to The Drum lists, chapters of the National Society of Black Engineers, National Assn. of Black Journalists, the Black Data Processing Assn., the 400 Black employees of Microsoft, and throughout the federal government, corporate America, and to small and large colleges and universities).

Anita Brown
@:-) Miss DC
Founder

We reach several thousand African Americans—several times a week. Consequently, membership is steadily rising. JOIN TODAY and be assured you'll be hooked up with this growing international movement!

To hell with the sites that Derrick engineered with his two degrees. Push aside the millions David made in the marketplace. Tone down the voices Barry promoted. Disregard the Afros that William and Tim Jenkins tried to make Go. Set aside the lineages Tyronne's BBS tried to trace back; the content that Lee made king; and the software that Kamal, Ken, and William turned black. Now to be fair, Anita would have said we needed all of them. And she would have been right.

For most of her life, Anita had been a secretary. But by the end of it she had become the architect of a movement still waiting to peak. Everybody had their own agenda, their own idea of what needed to happen, a plan for what would make all things right with the world. But it was Anita—someone with so little ego or money invested—who sat back, looked out onto the whole Internet world, and realized one, single-most important truth: a twenty-first-century revolution had to start with a revolution of technology.

Anita was the one who realized that black people needed to reposition ourselves not just to one another, to capital, or to the realities of white dominance. She realized that we must reposition ourselves to the technological tools that served to either support or resist white racial dominance. Anita realized that the revolution needed to be digitized and networked—not monetized—and that it must include citizens, government, educational institutions, and corporations.

She never quite articulated it this way. Anita recognized that we had seen some success. But she also realized that somehow we still needed to reformat the hardware, recode the software, overwrite black software's origin story. That story began to be written in 1960, the year that the computer revolution's vanguards at MIT abandoned their civil rights movement leaders organizing in the South.

That's when black software's code first began to be written. It was a program designed to isolate us, restrain us, contain us, lock us up.

Book Two

Chapter Eleven

COLLISION COURSE

No matter how far the Vanguard seemed to get, their progress was always constrained, prescribed, limited, like a dog on a chain that is only permitted to go so far before being choked by its collar.

The Vanguard could not see their chains. But they were there, placed there from the very beginning when the computer revolution first encountered the black freedom struggle. That story begins in 1960.

* * *

Adam Clayton Powell Jr. wasn't the first black member of Congress elected in the twentieth century. But he was the *first bad nigger* in Congress.[1] He fought Jim Crow. He fought poll taxes. He fought racist senators like South Carolina's segregationist Strom Thurmond. Powell was the son of mixed-raced parents. His slick, straight hair looked as tight as a Dapper Dan pomade model. And if someone thought his skin resembled theirs in the slightest, Powell made it abundantly clear that he didn't take shit from whitey—not on the streets of New York City, not in its city council, not on the floor of Congress.

Powell pastored Harlem's famed Abyssinian Baptist Church. He was blessed with the gift of speech, the power to reach each Negro on every street.[2] His silver tongue, black constituent base, and nearly two decades in Congress made Powell as formidable a foe as any lawmaker.

On July 14, 1961, Congressman Powell arrived at the Philadelphia, Pennsylvania, Sheraton Hotel. His flaming tongue and unmuted audacity were well intact. An NAACP meeting packed the hotel ballroom. Powell stoked the crowd's passions from the beginning. First, he reminded them that much of the civil rights battle up to that point had been waged in the courts, myopically focused on overthrowing Jim Crow. He said such *passive, piecemeal and paltry legalisms* were unfit for, and not enough to move, the cause forward fast enough. Powell reflected back just a few months into the past. He remembered the freedom rides. He conjured the sit-ins. This, he told the crowd, was the future. And, he congratulated those who participated for their success.

The struggle of the "sit-inners" and the Freedom Riders represents the basic break from legalisms. Their actions have stimulated and invigorated a democratic process. Their cause is in harmony with the principles of God. Their success is indispensable for the furtherance of our national interests before the congress of nations of the earth. Today, as a result, a catalytic agent has been injected into the House of America, so that the pretenders and the believers in Democracy are now clearly able to be counted and recorded. The piecemeal and tedious way of the courts has been thrust aside, and through the power of passive turbulence, a bridge is being created between Democracy's dream and Democracy's reality.

But Powell wasn't satisfied. Nor did he expect those gathered to be. They had to do more. Theirs had been a fight for the right to enjoy America's public accommodations. But Powell believed they needed to set their sights higher. The Negro should not be limited to just busting down mere social barriers. They had to confront power where power lived.

Therefore, tonight I call for the immediate acceleration of Phase Two; namely, economic freedom.

Powell held that crowd in the palm of his hands.

Amen!

Hallelujah!

Tell it!

Audience members waved one testifying hand in the air. The other flapped fans to cool themselves off. That room was hot!

Powell wanted them riled up. He aimed to instigate a fight. He prepared the people for political battle, to take their anger to the streets. But before he provided that call to action, Powell clarified the problem, as he saw it.

I shall not quote statistics. To do so would be a waste of your time and that of my staff. We know that the Afro-American is the last-hired, first-fired. We know that he pays a tax on being black, which makes him the lowest wage earner in this Nation. We know that he is quarantined, regardless of ability and education, so that his highest achievement can be the attainment of only creature comforts. We know that he composes the largest number of unemployed in this Nation today. We know that the new era of automation does not include him. We know that Government—local, State and Federal—rigidly excludes him or gives him token consideration at high levels and mass menial jobs at low levels.

The *new era of automation*—that's what the white boys at MIT so eagerly worked on. That is why they were so focused and could not be bothered by such inconsequential matters like ending racial discrimination.

Powell didn't have a PhD. He wasn't a scientist. He had probably never read *Cybernetics*, MIT professor Norbert Wiener's treatise on *Control and Communication in the Animal and the Machine*. But Powell also didn't come to Congress fresh from some back-country holler. His parents were wealthy. He had earned two degrees. He took a bachelors from Colgate College, and walked away from Columbia University with a masters degree in religious education.

Powell believed he did not need to read Wiener to understand why automation would revolutionize the American economy. He did not believe he needed to be in a lab designing new computing machines to fully grasp what automation meant for the Negro. He knew the US government had a firm hand in directing how automation would impact American citizens. Powell knew that the most important thing to understand at the time wasn't being taught at MIT. It was being lived by every black woman and man in America. Powell knew that America's racial order preceded Wiener's cybernetics by a hundred years—a century before the first IBM digital computer ever left the manufacturing floor, or landed in a MIT lab. Powell believed that America's racial politics, not the computer, would dictate automation's imperative and determine its impact.

But Powell was concerned. The technological revolution kicked into high gear. Black people stood to fall further down the economic ladder. But Powell had the ways and means to do something about it. Just that year he had seized the chairmanship of the powerful congressional Education and Labor Committee. *This* is why he whipped up that Philadelphia crowd. He did not just want to highlight the specter of automation for a country whose public and private sectors failed to live up to its ideals. Powell wanted to chastise and call the public and private sectors to account. Their failures had not afforded the Negro the freedom to compete with white people on a level playing field. They certainly were not prepared to compete with the automated machinery that threatened many of their livelihoods. Powell pushed immediate mass action.

We do here and now call for every American, with unparalleled militancy, to strike down the barriers of the economic blockade. We not only want to sit at the lunch counter, we want to work behind it! We not only want to ride the busses, we want to drive them![3]

Days after calling for action, Powell took action of his own. He introduced the Manpower and Development Training Act.[4] After a year of wrangling and revision, President John Kennedy ultimately signed into law the legislation Powell had helped to engineer.

* * *

Automation sat rather impatiently at the crossroads where civil rights and the computer revolution first crossed paths. The great labor leader, union president, and civil rights organizer A. Philip Randolph stood at that crossroads in 1962. There, he staged a conversation with himself (and with the hundreds of delegates gathered with him for the Fourth Triennial Convention, and Thirty-Seventh Anniversary of the Brotherhood of Sleeping Car Porters). He looked backward. He looked ahead. He reflected. And then, he spoke. He was both resolute, and uncertain.

Technological change cannot be stopped.
But the great masses of the people should not be required to bear the brunt of the impact of this great automation revolution, which is shaking the world.
The march of science cannot be arrested.
But what can be done about it?
You cannot destroy the machine, you cannot stifle the invention of various geniuses in the world.
Then what is to be done?

Powell thought he knew the way forward. He got John F. Kennedy to sign his prescriptions for advancing blacks in the new technological age into law, through the Manpower Development and Training Act of 1962. Then he hoped for the best. Randolph wasn't so sure he had any answers. For all they had done or proposed to do, both Randolph and Powell seemed—perhaps for once in their lives—unconfident.

They both stood at that crossroads together. They both saw the machinery of technological change barreling its way through, like a runaway train. But truth be told, neither of them fully appreciated what raced past their eyes. They saw the computer revolution for what it was—what it was in 1962, that is. Forget predicting. They could not even imagine what impact the computer revolution would have on black futures. But how could they? And how could everyday black folk? Many of them didn't have the luxury to prognosticate or imagine, or dare to dream. They were so far removed from the computing systems that were, at that very moment, remaking the world.

* * *

Adam Clayton Powell Jr., A. Philip Randolph, Martin Luther King Jr., and Bayard Rustin openly lamented what they thought automation would bring to black America. They also warned about the dangers of "cybernation." But for each of them, automation and cybernation occupied one side of a single coin. Dr. Donald N. Michael helped other civil rights activists, professors, technologists, policy makers, and planners to understand. But there was a very real difference between automation and cybernation, and this difference made the outlook for black America even more bleak.

Dr. Michael earned his first degree, in physics, from Harvard; his second, in sociology, from the University of Chicago. He returned to Harvard to get a PhD in social psychology. In 1962 he was the director of planning and research for the Peace Research Institute, located in Washington, DC. During this time, he took an interest in understanding how the new computer revolution would impact every aspect of society.

Michael's book *Cybernation: The Silent Conquest*[5] started its journey right there at that same crossroads where Powell and Randolph stood contemplating automation's impact on the Negro. Imagine the conversation that *might have* taken place had they actually been allowed to meet face-to-face at that crucial intersection: congressman to union leader to technology planner. Michael could help them see what they had not, and could not.

Powell's and Randolph's concerns about automation were based on what they lived and knew about black people's past in America. This was something that Michael could never fully appreciate. But Michael knew technology. Powell and Randolph did not have Michael's depth of knowledge in that area, but all of them knew very well how government and private industry worked.

Technological change cannot be stopped.

Yes, you are right.

We know the era of automation does not include the Negro.

Well, at least not in the way you think.

Then what is to be done?

First you must be willing to consider a possibility that you may have never entertained.

Powell and Randolph were reasonable and dignified gentlemen. They enjoyed a good debate. They were always happy to entertain a new perspective.

Please, continue.

There is a very good possibility that automation is so different in degree as to be a profound difference in kind, and that will pose unique problems for society, challenging our basic values and the ways in which we express and enforce them.

So, what are we missing? What do we need to know in order to understand what automation means for the Negro?

We have to know something of the nature and use of automation and computers.

We know what automation is.

I know you very well do. Your work in industry and government makes you familiar with the fact that automation is made up of devices that automatically perform sensing and motor tasks, replacing or improving on human capacities for performing these functions.

Yes.

But computers do something a little bit different. The computer is composed of devices that perform, very rapidly, routine or complex logical and decision-making tasks, replacing or improving on human capacities for performing these functions.

So if Adam and I wanted the building trades to be automated, that would not mean inventing machines to do the various tasks now done by black men and women?

Yes, you are right.

Rather, buildings would be redesigned so that they could be built by machines?

Correct. One might invent an automatic bricklayer, but it is more likely that housing would be designed so that bricks would not need to be laid.

We see. So a black man or woman could still find a good job in the housing industry if she or he knows how to design homes that did not need bricks?

Precisely.

But, you do understand…

Yes, I know. I know. There is an increasingly lopsided Negro-to-white unemployment ratio as the dock, factory, and mine operations where Negroes have hitherto found their steadiest employment are cybernated. This, plus the handicaps of bias in hiring and lack of educational opportunity, leaves Negroes very few chances.…

Exactly. Then what is to be done?

Michael would not have been able to provide an answer to their questions at that juncture of the conversation—had it happened and proceeded as imagined. Michael still had yet to fully explain what cybernation meant.

Had he continued, Michael would have simplified things a bit further for them, in the same way he would have had to for 9.9 out of 10 other US citizens. "Cybernation" referred to the mechanized production of goods, aided by a computer. It included the automated processing of data. Cybernation referred to the systemic and regulatory communication and control relationship between human beings and computers. A "cybernated" system—again, simply put—included a production process governed by a computer. The machine can process production data. It could "learn" from that data (i.e., memorize, interpret, and analyze). It could use what it "learned" to regulate a continuous production process with minimal human intervention.[6]

Putting on his social and institutional planning hat, Michael further would have pointed out the two levels at which human beings engage with cybernated systems. First, there are programmers. They encode the logic, and write the instructions (the algorithm, the program, the software) that the computer will process. The programmer also creates the instructions the computer will use to copy and store both its inputs (software) and outputs (data) onto magnetic tape. The second type of human agent that engages a cybernated system is the technician. The technician merely inputs the required software from the tape on which it was stored, into the computer.

The computer then executes the pre-programmed instructions. The technician makes sure the machine continues to work as designed. The technician troubleshoots. The technician helps other users make the computer do what it is supposed to do. But, the computer technician does not fundamentally alter the system's hardware, software, or the system itself.

Michael's crude example also outlined an equally as crude, but useful, theory. It explains how cybernation produces value. Further, it explains who has, and how one gains, power in a technological (computational) society. Cybernation made decision-making easier and more efficient. Private industry benefited; it increased the speed and scale of production. In turn it increased revenue. Governments benefited; it facilitated more efficient, mass decision-making. Cybernation helped government better manage its citizens. It made them easier to govern, regulate, and control. Cybernation made it easier to design and engineer a society according to the goals and interests of the state.

Michaels explained that one's relationship and proximity to technology—computing systems—determines one's technological power. As an example, let's place the computer system at the top of a pyramid. Just below it are private industry executives who own the system. They control its manufacture and distribution. Heads of state and those leading government agencies and institutions stand in very close proximity. They can exert influence over private industry by providing incentives or disincentives to comply with governmental prerogatives. We call this regulation.

Overlapping industry executives and heads of state, and sitting somewhere in between or just below, are those who design computer systems. This includes a range of people in occupations such as research scientists, mathematicians, designers, hardware and software developers, engineers, statisticians, research analysts, and the like. Toward the bottom lie your technicians. Then users. Then those with little or no connection to computers.

Adam Clayton Powell and A. Philip Randolph both wanted the same thing. They wanted to see and make a way for the Negro to access, achieve, and attain economic security, freedom, and power. What they had not fully realized was how much that vision would be determined by the Negro's relationship to computing, and a new form of technological power. Had they, and those like them, had the opportunity to engage in that imagined conversation with Donald Michael, they might have been able to more adroitly plot and plan a way forward.

But Randolph and Powell were already approaching the ends of their careers, and their lives. And neither the Union Hall nor the Congress was the best place to learn about, discuss, and take the time to plan the future. Doing so in their twenties would have been a better time. A higher education institution, especially an elite school of science and engineering that was embedded in both industry and government, would have been a more appropriate place. But it was already too late. And, apparently, institutions like this did not quite welcome Negroes.

* * *

On Friday, August 6, 1965, President Lyndon Johnson stood at a Capitol rotunda podium, poised to sign the most far-reaching and impactful civil rights legislation in US history. Civil rights leaders, from Rosa Parks, to Rev. Dr. Martin Luther King Jr., Rev. Ralph David Abernathy, and others, looked on. They smiled triumphantly. They were hopeful about the future. The president signed the Voting Rights Act into law. By stroke of a presidential pen, black people had seized and secured the greatest power ever at their disposal: the franchise.

Just four days earlier, however, the US Department of Labor began the week churning out a relatively uncelebrated news release. The weekly news digest regularly filled itself with the problem of Negro employment. The August 2 edition, authored by US Labor Secretary W. Willard Wirtz, lead with the headline: "Heads-on Collision Course for Civil Rights and Automation." The news release proffered two immutable truths.

Call it automation, cybernation or age of the robots, the technological revolution is here and here to stay; call it civil rights, or equal opportunity or "peaceable protest for freedom now," it is equally evident that the Negro revolution is here.

These lines faded into the long shadow cast by the end of the week's voting rights victory. Johnson was privileged to pay in full one of America's long overdue promissory notes. But his labor secretary ended his weekly memo with this cryptic reality check:

There are machines now which can play excellent games of checkers, and they can play pretty good games of chess; they can play a fair hand of bridge; they can interpret books. They are doing in a good many ways skilled as well as unskilled jobs in the economy. They have no hands but they can tabulate checks, and they make no mistakes. They have no eyes and yet they watch over the industrial process without ever blinking. It has taken only thirty years to move from the fantasy of "Rossum's Universal Robot" to the reality of Univac and its brother computers, and today's reality is more fantastic than the fiction of the 1920's.

To some, Wirtz's words read like a primer on the days' advances in computerization and artificial intelligence. But they were not. Wirtz offered a prescient declaration about where the Negro stood in comparison to the times' "thinking" machines.

When the US government predicted the collision between the civil rights revolution and the computer revolution, it did not imagine a fender bender that left both equally damaged. It imagined a collision that left the computer revolution unscathed, and the civil rights revolution twisted and mangled up within it. Why did this collision and its predicted outcome seem so inevitable?

Chapter Twelve

THE REVOLUTION, BROUGHT
TO YOU BY IBM

On July 23, 1965, President Lyndon Johnson issued Executive Order 11236. The order established the President's Commission on Law Enforcement and the Administration of Justice. Johnson just called the group The Commission. It had twenty members, and the president appointed them all himself, including its chairman, the Attorney General of the United States Nicholas Katzenbach. The order mandated that the commission *inquire into the causes of crime and delinquency, measures for their prevention, the adequacy of law enforcement and administration of justice, and the factors encouraging respect or disrespect for law, at the national, State, and local levels, and make such studies, conduct such hearings, and request such information as it deems appropriate for this purpose.*

Katzenbach organized The Commission's investigative work by forming separate, independent task forces. He charged each to focus on one or more crime and law enforcement matters. One task force focused on police, public safety, and narcotics. One assessed the crime problem. Another targeted organized crime, corrections, and the courts. There were others. But the president and Katzenbach invested a considerable stake in The Commission's Task Force on Science and Technology.

* * *

Fall came, and Watts still simmered. Nearly fifty years of Negro silence had stripped away its very soul.[1] *Tired*, as one man put it, *of being pushed around by you white people*, Watts finally woke. The sprawling urban community had confined three-quarters of Los Angeles's black residents. For six long, sweltering August days in 1965, it raised its voice. That voice reverberated against a cacophonous torrent of rocks and gunfire, broken windows and Molotov cocktails. As one, giant, clenched, and raised black fist, Watts rose. It raised itself up to become a towering national beacon for a newfound black power.

But by the time the LAPD and 14,000 National Guardsmen ended their six-day, fifty-four-square-block siege, thirty-four souls—all of them black—lay bloodied and still. Another thousand nursed injuries. Four thousand shuffled their way, handcuffed, through police car and jail cell revolving doors. Forty million dollars' worth of buildings, businesses, automobiles, and other neighborhood property smoldered in ashes.

White Los Angeles searched for explanations. Parochial and paternalistic, they thought their blacks had it better than any in the country, especially compared to the South. Watts caught them off guard. Watts's sudden resistance required an answer. Public officials, professors, and newsmen produced many of them. All exploited Los Angeles's and the American public's rapt attention. Two weeks after the uprising ended, California governor Pat Brown appointed a commission of his own. He tapped former CIA director Roy McCone to lead the group. Brown tasked the group to develop a *comprehensive and detailed chronology and description of the disorders*. The governor also charged the group to propose steps that government, private enterprise, and citizens could take to ensure Watts never happened again.

McCone's commission delivered its 101-page report on December 2, 1965. The report's title, *Violence in the City*, said it all. The title defined the situation as the commission saw it. Watts connected black Los Angeles's natural tendency toward crime and violence to white Los Angeles's innate fears. The report downplayed how many "rioters" participated. It said it was just the riffraff. New York governor Nelson Rockefeller, and the New York City mayor used the same explanation to explain away the Harlem and Rochester riots that rocked their respective cities in the summer of 1964. Long-standing and pervasive police brutality had fueled Watts's frustrations. But the commission's report focused on education and work. Negroes had had too little of each, for too long. That's why they rioted, the report explained. They were equipped to do nothing else.

But just as the commission ended its work, two political science professors delved deeper, and uncovered a contradictory story. According to the professors—one from the University of California, Los Angeles, the other from Yale University—almost 40 percent of Watts's black population participated in the uprising. Thirty-five thousand adults participated directly. Others observed and supported those who participated. Three out of every four thought the days-long rebellion yielded a favorable outcome. The professors said the revolt touched the entire black community. They debunked the commission's so-called riffraff theory. And, members of the community refused to define the week's actors and actions as deviant.

The contrasting reports were compelling, at least to the kinds of people who read such things. But neither could compete with the number of eyeballs that CBS's Special Report drew from across Los Angeles and throughout the nation. On December 7, 1965, the network debuted "Watts: Riot or Revolt."

* * *

A rapidly crescendoing bugle call opened the broadcast, sounding like the start from one of Walt Disney's war propaganda films. The words *A Presentation of CBS News* sat simply, within a small, white frame, just left of center on the otherwise black television screen. As its revelry peaked, the frame faded to black, leaving just a three-dimensional logo. *CBS NEWS.*

The film opened right in the middle of a nighttime street scene. One tall street lamp exposed a light, hovering fog. It illuminated tense confrontation. A white cop stood near a parked vehicle's hood, his rifle trained on a black man standing at the car's rear. The black man stood silent, hands hoisted high into the air. *Hands up high,* the officer commanded. *You can lift 'em higher 'an that,* he shouted, exposing a tinge of sarcasm. *First one drops their hands is a dead man.*

Steady gun pops broke the silence. They're off in the background, but they let the man know the officer's threat is real. The screen rapidly cut to successive moving images. White cops fired rifles and semiautomatic weapons into the darkness. Bursts of what sounded like cannon fire accompanied them.

Then, a television newsman's voice broke in. The voice emulated Rod Serling's style—prescient, authoritative, every word carefully enunciated as if each syllable were pregnant with meaning.

An absolutely incredible scene. A gun battle, in the middle of Broadway ... the streets of Los Angeles. A small army of policemen, most of them carrying shotguns. National guardsmen, riding jeeps with thirty-caliber machine guns. Bodies of several Negroes who have been shot already in this battle, stretched out beside the curb. Acres and acres of broken glass. Burned, looted stores, offices. And now this hunt. Like something out of a bad war movie. A Western, perhaps. Policemen on the rooftops, in the streets. They think now they've found the men they were after.

After the dramatic teaser, KNXT-Los Angeles news correspondent Bill Stout took over. He narrated the unfolding story as images paraded across the screen. Voices teased upcoming statements by some of the drama's key players.

I believe it started 400 years ago, one man says, while another explains that "rioting" was their community's only recourse. *Rather than listen, the police just want to beat my brains out.*

The next scene cut quickly to a government meeting room. *We cannot, Governor ... uhh ... tell you any one particular reason why the riots took place.* McCone spoke with a slight grin. They were perfunctory words; the governor knew full well what the report said.

Stout provided context and gave his audience a simple, neat, tidy, and impenetrable frame, before they immersed themselves in the drama.

Tonight's CBS Reports. The question of Watts. Was it a local riot, or the beginning of a national revolt? What started it? What stopped it? Will there be another Watts? John McCone has just presented Governor Edmund Brown of California his investigating committee's report seeking answers to just such questions. These findings are an integral part of what follows, the CBS Reports study of the principal events and causes of the nightmare in Watts.

Titillating, anxiety provoking, powerful, *Watts: Riot or Revolt* ran for close to an hour. It would be the first time the computer industry, represented by its market leader, had put its interests in the so-called race problem on full display for the nation, and the world, to witness.

* * *

Cymbals crashed, practically cutting off narrator Bill Stout's final introductory words. Then, more revelry, while television cameras guided viewers slowly through Watts's destruction. A bombed-out shopping center reduced to iron and rubble. Papers, sheetrock, and concrete were strewn about as if they were the remnants of World War II. Entryways gaped where walls once stood. Iron frames affixed to rooftops exposed rust once covered by business signs.

Then, quite literally out of the blue, three letters emerge from the wreckage. Swelling in size as they seized more of the screen's territory, International Business Machine's trademarked logo triumphantly occupied the screen's center. A voice announced, *International Business Machines. IBM Presents, CBS Reports: Watts, Riot or Revolt.*

No one would have put the two, IBM and Watts, together. One towered over its neighbors among the buildings lining Wilshire Blvd. It sat just a short hop from Hollywood and Beverly Hills. The other lay a considerable distance to the south on Interstate 10. Professionals, managers, bookkeepers, typists, clerks, duplicating machine and computer operators took up residence in one. But not more than about a hundred such folks could be found in the other. One surrounded itself with innovation, resources, and privilege. The other seemed to have accepted that theirs was a ghetto slum. One was almost entirely white, the other black, and those who lived in each scarcely walked the other's streets.

So what brought these two starkly contrasted worlds together in the summer of 1965? More pointedly, what was IBM doing in the middle of the Watts riots? Television facilitated this uncharacteristic moment of collective, urban miscegenation. The experience, for the millions of viewers captivated by the new technology's magic, seemed as real as the revolution itself.

* * *

In 1965, IBM raced—against no one but itself really—to dominate a burgeoning computing industry. Headquartered in Armonk, New York, the company, in 1965 alone, built or acquired twenty-eight new facilities. Three million square feet of manufacturing space sprawled across the globe, from East Fishkill, New York; north to Toronto, Canada; south to Raleigh, North Carolina; east to Vimercate, Italy; and west to San Jose, California. IBM, together with its international subsidiary, World Trade Corporation, boasted a global footprint. It had accumulated more than $3 billion in total assets, took in nearly a half billion dollars in profits, increased its stock price to $6 per share, mushroomed to 172,445 employees, and, it ranked among the top ten companies on Fortune Magazine's new top 500 list of US companies.[2] There wasn't a competitor in sight, and IBM had been dodging antitrust lawsuits since 1952.

IBM's chairman, Thomas J. Watson Jr., reported to the company's share-holders in a letter that introduced the year's annual report. He described that the company had *conducted the most intensive engineering, manufacturing, and marketing effort in history* to launch its most significant product in its history to that point. The IBM System/360 featured the company's groundbreaking solid-state technology. It produced a smaller, faster, more flexible, and versatile family of computer systems. They were built to meet what IBM described as *unprecedented demand* spanning the wide field of data processing. The cover of that year's annual report captured the achievement's significance. There, a fishbowl view of its Poughkeepsie, New York, "programming center" filled the page. It was the *largest commercial center in the world*. The document explained that the facility was part of a *worldwide network of IBM centers developing the step-by-step programs of instruction which enable System/360 to solve a vast range of complex problems.*

IBM's historical corporate entanglements put them in Southern California as early as 1916. But it wasn't until 1958 that the company established a more prominent foothold in the region. That year IBM built a 234,000-square-foot office building for its Western Regional Headquarters, just west of downtown Los Angeles. It was fifteen miles, a ninety-minute drive, north-west from Watts. The office housed IBM's Data Processing Division. The hub processed business and scientific data for customers in twelve surrounding states, as well as Hawaii and Alaska.

* * *

The explanation could be simple. IBM was a billion-dollar corporation. The airwaves were made for advertising, and "Watts: Riot or Revolt" was just good television. From the dawn of commercial TV in 1941, CBS actively experimented with the technology's capabilities. It launched the CBS Television Audience Research Institute in 1944 to document its observations.[3] It studied television's visual appeal and deciphered why certain television programs attracted certain audiences and not others. It scrutinized the voice qualities and body features that produced an authoritative news anchor.

When CBS aired its first sponsored television program in 1946, the broadcaster narrowed its research focus. It wanted to assess whether a specific new format—television news—could thrive. It already knew that television had great power to communicate through visual images. It used newsreel footage, still photographs, and animated maps to manufacture and communicate a sense of gravity and importance to newsworthy events.[4]

CBS also benefited at the time from a new venture. It partnered with former advertising executives Philip McHugh and Peter Hoffman, and University of Chicago anthropologist Lloyd Warner. Just a few years earlier, Warner had formed Social Research Incorporated (SRI). Warner convinced McHugh and Hoffman to join. As the first television news "consultants," McHugh's and Hoffman's role would be to sell research to broadcasters. They started by teaming up with four SRI sociologists to conduct the first audience research on television news.[5]

They discovered that television was a "class conflicted medium." Television in general particularly appealed to the middle and lower classes. The medium provided content and characters with whom they could identify. But television news in particular attracted those at society's higher echelons. These country club elites favored programming that mirrored themselves and the worlds they lived in.

But opportunity lay between the clashing perspectives about television news. Middle- and lower-class people liked television news when it featured relevant content about their local living and working environment. Highly visual and dramatic content appealed to them especially. This made weather-related news appealing; it was useful. It also made crime news alluring; it was dramatic.

Television news' visual and dramatic appeal also had much to do with ushering the black masses to television in the early 1960s. Blacks shifted from being low to high media users practically overnight. Why? Because television featured visual, local, dramatic and—most of all—relevant content about civil rights.[6]

In any case, KNXT wanted to produce more effective programming. It wanted to garner higher ratings. It really wanted to increase advertising revenue. KNXT did not know how to do it on its own, so it called on SRI. Los Angeles's television market was, like television itself, conflicted, with audiences split across the same class dividing lines.

CBS had already received a national Peabody Award for the kind of news programming it began to produce.

In a year when the intelligent, adult television audience has been consistently short-changed by networks wooing teenagers, CBS Reports stands out as a heartening exception to the trend. Judicious selection of material, first-rate editing, and superb production have been the hallmarks of every CBS Reports program.[7]

But when SRI's McHugh canvassed mundane corridors beyond the Hollywood Hills, everyday people were quite emphatic. Television newsmen were *over their heads* and *conceited*. They trained their cameras on people and surroundings far removed from everyday people's realities. The masses wanted action, not droning chatter. They want to see life as they experienced it—moving, not standing still like some out-of-touch podium-perched politician peppering the public with highfalutin words. The masses wanted the cameras pointed where they lived. On their streets. On their people. McHugh delivered this news to KNXT, and strongly recommended the station follow its precise instructions.

One month later Watts ignited. KNXT and Los Angeles's viewers got what they wanted. KNXT had heeded its consultant's advice, down to the word. And Watts felt like a gifted opportunity to experiment.

* * *

Chaos erupted, and CBS cameramen stormed Watts's wide, suburban-like streets like it was D-Day. Bill Stout had been a newspaper man. He had written for the *Los Angeles Times* before he joined CBS. Stout covered Watts on nightly newscasts throughout the six days' upheaval. He had a face for television, a voice for radio, and a seemingly authoritative command over the facts of the city—its people, its players, its politics. He fit CBS's formula for success. And, like the newsmen of his day, he assumed two roles—reporter and commentator. He brought stories to the public, and he distilled meaning for them.

Watts: Riot or Revolt seemed to be designed and engineered straight from the CBS Audience Institute and SRI blueprint. The fifty-minute broadcast provided "eyewitness" accounts directly from Watts's burning streets. LAPD officers marched hundreds deep. Fourteen-thousand National Guard troops parachuted in for extra muscle and firepower. Watts's black citizens, gathering on street corners and in churches and meeting halls, tried to explain their actions. A CBS newsman to make sense of it all for the American viewing public completed the cast.

Stout put his official sources in front of his viewers. An aloof and (literally) distant Mayor Sam Yorty made comments for cameras during layovers between flights to previously scheduled speaking engagements. Police Chief William Parker, a former military officer and lawyer, upheld orders both lawful and racial. In the mayor's absence, he exercised virtually full control over the city. Dr. Daniel Patrick Moynihan also graced the cameras to offer much-needed solutions to the array of urban problems on display in Watts throughout that week. Moynihan was a sociologist and researcher at the MIT-Harvard Joint Center for Urban Studies. He was an expert on urban conditions and the underclass black folk who lived in them. He was a labor secretary and apologist for President Lyndon Johnson's War on Poverty.

The broadcast featured a crime drama on a scale no McHugh focus group could have ever imagined. And CBS crafted a dramatic narrative for the public, one full of people they could love and hate, identify with and fear.

Violence produced all this. The burning and looting, the shooting and beating went on for nearly a week. The evils which can be endured with patience as long as they are inevitable, seem intolerable as soon as hope can be entertained of escaping them.

Stout's quote errs slightly. Nonetheless, he skipped over decades of South Los Angeles's history of forced segregation, brutal real estate redlining, inadequate transportation, and staggering police brutality.

In the past twenty-five years, more than 300,000 Negroes from other parts of the United States have come to Los Angeles in the hopes of escaping evils they had endured with patience. But on the night of Wednesday, August the 11th, that patience ran out.

Fire lit up a television screen darkened by nightfall. Dissonant sirens invaded, unnerving any viewer. Rapidly edited scenes panned the chaos while black voices pierced the silence.

KILL THE WHITE MAN
Kill the white man.
Get 'em!

In the background, a sinister voice chuckled. It exuded a kind of murderous release.

Get the white man.
Kill the white man!

Stout's voice broke through the noise.

It was the most widespread, most destructive racial violence in American history. White people driving through the riot area were considered fair game, whether young or old, men or women. Their cars were battered, their drivers stoned and beaten—and the cars were burned.

CBS had produced an American race war, practically live for Los Angeles's television audience. The nation's most flagrant fears played out on the screen even more powerfully than they once had when *Birth of a Nation* pictured a similar narrative fifty years before.

Yes. IBM could have been simply hitching its wagons to what *Watts: Riot or Revolt* represented, a public demonstration of the very best the new television technology had to offer. Yes, IBM, and a small handful of similarly dominant corporations, routinely dabbled in this kind of sponsor-driven brand association.

But perhaps IBM wanted to do more than associate its corporate brand with the broadcast. Maybe it intended its sponsorship to communicate something different. After all, the company had spent millions being very intentional about how the industry and the public perceived it. Perhaps IBM's corporate eyes viewed Watts the same way Stout's viewers were seeing it depicted on the screen.

* * *

When *Watts: Riot or Revolt* aired, the single-sponsor era, and the mantra *he who pays the piper calls the tune*, still prevailed.[8] In fact, it would not have been out of the ordinary at the time for a corporation like IBM not only to pay the piper and call the tune, but to have proposed the idea in the first place, a practice it started back when the company first sponsored the airing of the 1960 Olympics.[9] That is, IBM may have very well initiated the broadcast's production.

Maybe IBM pushed CBS to produce *Watts: Riot or Revolt*. Maybe affixing its corporate mark to the documentary meant IBM agreed with its message. If so, what message about Watts—its people, its actions, and the city, state, and nation's response—did IBM align itself with?

Stout's news report appeared to balance opposing explanations about what happened in Watts and its significance. But rather than position himself as fulcrum, balancing the weighted opinions advanced by both sides, newsman Bill Stout did more than tip the scales to the side where power already resided.

Stout's voice led the way. He marshaled statistics to paint a simple picture. Few of the rioters had graduated from high school. Many of them were illiterate. A third of the adults were unemployed; most on welfare. Most of the kids were dropouts, from broken homes.

Most residents are newcomers, who joined the modern gold rush to California over the past twenty-five years. Many are newcomers from the most backward parts of the deep South—poor and ignorant Negroes who have no skills to offer a big city employer. No desire for classroom learning, not even the knowledge of how to live in urban surroundings, often not even the knowledge of how to use plumbing. They crowd together, these back country refugees, a thousand new ones every month piling into Los Angeles, and they find in the land of golden promise that there are still white lawmen, white merchants, white landlords.

Did these poor fools think they were going to escape American white supremacy's long reach when they moved out West? Stout seemed to think they did. He argued that people's naive sensibility that they could escape white supremacy's stranglehold is what fueled their frustration. That, he said, was the fuel that lit the spark that ignited the city.

Chief Parker entered the documentary next. He framed Watts's turmoil just as Stout had. Parker, however, took aim at civil rights leaders.

They're constantly trying to reach these groups for political balance of power by catering to their emotions. You're dislocated. You're abused . . . because of your color . . . your progenitors were oppressed. You haven't been given the share of materialistic things you're entitled to.

Parker was skinny. His eyes drooped, and his jowls sagged. His crew cut looked like it remained untouched from his military days. He had a legitimate slur. And if you put aside that he lived in Southern California, your ears could catch the accent of a slight, Southern drawl. He didn't spread southern justice like Alabama's Bull Connor. He exuded the same vehemence, but spared his fire hose.

Parker made it plain to Stout.

Many of these people, in our present system are not in a position to set into the economy and become affluent unless somebody hands it to them.

Parker also reminded Stout that civil disobedience erodes respect for law and order. Sitting on the other side of Parker's desk, Stout teed Parker up for another swing at Los Angeles's rioting Negro population.

Do you think that what happened was simply a racial manifestation of disrespect for law, or do you see it as something related to the social and economic strivings of the Negro?

I think all those factors are involved.

In creating the situation, where was the failure? The city? The county? The schools?

Indignant and practically rising from his seat, Parker launched in.

This is one of the difficulties in meeting this, it's that we're trying to find a failure, other than the people themselves.

Agitated, he continued to set the record straight.

They came in and flooded a community that wasn't prepared to meet them....We didn't ask these people to come here, and suddenly they want our total community to adjust itself to...a small segment that has suddenly come in and taken over a section. I think this is unreasonable. I think that we're almost sadistic in the way that we're trying to punish ourselves over this thing without realizing what we have destroyed, which is a sense of responsibility for our own actions.

Stout did nothing to push back against Chief Parker. Rather than challenge any of Parker's explanations or premises, he insisted that *most whites share Parker's views.*

Hard-nosed and convincing as Parker came across to Los Angeles's whites, Secretary Moynihan got to deliver the expert explanations for viewers. After all, Moynihan's words earlier that year had been prophetic. Just five months before Watts erupted, Moynihan had completed and distributed his new report, *The Negro Family: The Case for National Action.*

The effort, no matter how savage and brutal, of some State and local governments to thwart the exercise of those rights is doomed. The nation will not put up with it—least of all the Negroes.

Moynihan strayed from Parker and Stout's definition of the situation. He acknowledged that the Negro's position followed America's slave trading legacy. Despite his departure from Stout and Parker, though, Moynihan still ended up at the same place.

The evidence—not final, but powerfully persuasive—is that the Negro family in the urban ghettos is crumbling. A middle-class group has managed to save itself, but for vast numbers of the unskilled, poorly educated, city working class the fabric of conventional social relationships has all but disintegrated, he pronounced. *So long as this situation persists, the cycle of poverty and disadvantage will continue to repeat itself.* These words introduced Moynihan's seventy-five-page solution for halting the cycle.

A slaphappy Moynihan smacked his desk as he offered his solution to the problem.

We got to get men to work! A man can't run his family if he doesn't have a job. . . . Creating jobs for men is no secret. We know how to do it. We've just gotta get it clear in our minds, either we do it, or we're gonna spoil this beautiful country of ours—and that means spoiling those pretty white suburbs just as much as spoiling those nasty and ugly places like Watts.

Stout closed *Watts: Riot or Revolt* with a pronouncement, taken from the McCone Commission report.

What shall it avail our nation if we can place a man on the moon, but cannot cure the sickness in our cities?

A phone rang. The revelry ramped up once again. White lines stretched across the screen before they coalesced into IBM's logo.

CBS Reports has been brought to you by International Business Machines—I. B. M.

* * *

Five months before Watts erupted, President Johnson had declared, *There is no Negro problem. There is no Southern problem. There is no Northern problem. There is only an American problem. And we are met here tonight as Americans— not as Democrats or Republicans—we are met here as Americans to solve that problem.*[10]

Four months after Watts, IBM's Watson reflected on the year. Who knows if he thought much about that month in August when Watts exploded. Who knows if he approved IBM's sponsorship or not—or if he even knew about it.

Nevertheless, Watson looked back on the year. He contemplated the billion dollars the company earned. He reflected on the reason that IBM exists. The secret to its success. Its sole business model. Its mission.

The company's mission is to solve problems as they may exist in business, science, defense, education, medicine, space exploration and other areas. To perform this mission, the company conducts basic and applied research, development and manufacturing engineering, marketing and maintenance operations.

As far as the nation was concerned, its national problem and its Negro problem became one and the same. And the problem with perpetually being a problem is that you become the object of anything and everything that calls itself a solution—including a computer.

Watts and black America needed to be processed, brought in line with the rest of the country. America needed to find a way to reincorporate black people into a social order that made sense, and where everyone stayed in their rightful place.

Even in 1965 IBM had been no stranger to such a problem. It knew how to engineer national order. Or at least it had started to try. The same year that IBM inserted itself into Watts's racial affairs, the company had bid on a project on the other side of the world. It proposed to build a computerized system to identify and track black South Africans. The country's apartheid government ultimately awarded the contract to another company. Not to worry. IBM would later get to operationalize the system it had already invested in building.

IBM developed what was called the "Book of Life."[11] This "book" was a comprehensive and computerized identity register. White South Africans would use it in nefarious ways to prop up apartheid by, among other things, stripping black South Africans of their citizenship and ways of making a living. But the Book of Life was different from a census. A census enumerates and then aggregates its citizens for processing. The Book of Life provided the means to account for each and every individual person, whose identity was fixed in the record. This was a surveillance mechanism that would allow the government to control individual black South Africans' movements in and beyond the country's borders. It could have allowed that same government to monitor the size and expansion of its black population and act accordingly if they believed that blacks began to threaten the apartheid government. This was not dissimilar to another connection between IBM and attempts further back in history to control population during the Holocaust. The company, through various subsidiaries, had developed the tabulating machines card system used to identify, count, sort, and track Jews and other undesirables.[12]

But well beyond these excursions, IBM had begun to solidify its brand as the company that could identify and find computational, automated solutions to the world's most challenging problems. Taming the natural environment or taming the political problems spurred by a backward race? IBM made no distinction.

That is one of the reasons that not long after Watts, the US government once again called IBM.

Chapter Thirteen

THE COMMITTEEMEN

In walked the Committeemen. One by one they took their seats at the table. President Johnson had promised the American people that he would do something about the nation's growing crime problem, though he was not particularly convinced there was one. But the people had forced his hand. Fear—both real and imagined—gripped their minds. They saw it when they walked their neighborhood streets. They felt vulnerable when they ventured into their cities. Black people had the least police protection from crime. But who are we kidding? White people pushed the idea that crime had infected every corner of America. More and more they saw black people when they conjured the nation's crime problem. And Watts had given them the latest reason to be afraid.

West Coast officials focused on Watts through the end of 1965. Meanwhile, the Committeemen huddled in Alexandria, Virginia. They gathered at the Institute for Defense Analyses, ready to fulfill their national duty. Their science and technology had fended off foreign aggressors. Now they would set their minds to working on facets of our ever-irritating race question. That question, of course, was masked as a problem of crime, law enforcement and the administration of justice.

Attorney General Nicholas Katzenbach tasked the Committeemen— members of the Commission's Science and Technology Task Force—to inject scientific insight into law enforcement challenges. Katzenbach asked them to determine what science and technology could contribute to solving criminal justice problems. The Committeemen—scientists, advisers, consultants, and staff seated around the table—were hand selected by the president and Katzenbach. They represented the day's technological vanguard. They were well situated in their scientific communities, well positioned at their academic institutions, and well placed in industry and in government.

Three teams composed the task force. The Honorable Charles D. Breitel, an associate judge for the US Court of Appeals in New York, led team one. The president appointed the members of this team. Its members included Dr. James Fletcher, president of the University of Utah; LAPD Police Chief Thomas Reddin; Dr. Robert Sproull, vice president for academic affairs, Cornell University; Professor Adam Yarmolinsky from Harvard Law School; James Q. Wilson, Harvard government professor and fellow at MIT's Joint Center for Urban Studies; Dr. David Robinson, President Johnson's science adviser; and Dr. Eugene Fubini, a vice president at IBM. In the coming days, team one would investigate, review research findings, deliberate, and deliver a final report to Katzenbach.

But a scientist, Dr. Alfred Blumstein, directed team two. That team was responsible for the Committeemen's research and technical work. This is what lay at the heart of the task force's efforts. Blumstein was an engineer, physicist, and operations researcher. He stacked his team with research and technical advisors. Each had been steeped for a decade in government and military science and technology work. Chief among them was Dr. Saul I. Gass, an operations researcher. It was a field he had pioneered—first as an air force mathematician, then, as an applied researcher at IBM. Gass had presided over Project Mercury, which sent the first human flying into outer space. When he joined the Committeemen, Gass was managing the entirety of IBM's federally contracted civil programs. A group from the Institute for Defense Analyses and two electronics consultants rounded out the team.

Finally, team three engaged a separate group, consisting of eighteen advisors. These advisors conducted specialized research projects that were independent of, but coordinated and closely monitored by, all of the Committeemen. Three members of this team held scientific and sales positions at IBM.

Katzenbach explained the reasoning—though a bit opaquely—behind the task force's organization.

Since the task force staff of scientists and engineers had little prior knowledge of criminal justice operations and problems, it relied heavily on the Crime Commission staff and numerous criminal justice officials for identification of the operational problems of the system.... The Federal Bureau of Investigation was extremely helpful in this regard, especially in providing data on crime in the United States. Among State agencies, the California Bureau of Criminal Statistics and the New York State Identification and Intelligence System also made valuable contributions.

* * *

R. E. McDonnell sauntered into the office like he belonged there. And, Mr. Henry S. Ruth Jr. thought nothing of his casually dropping by. McDonnell's feet had wandered these 1.2 million square feet of Pennsylvania Avenue real estate before, hawking his technological wares. He had particularly spent a lot of time roaming the FBI's offices. This was the active perch from which J. Edgar Hoover directed his counterintelligence program. It was the place where he enlisted its own agents, and army intelligence officers, to covertly spy on civil rights leaders and other feared national threats.

But sometime on or around Thursday, November 18, 1965, McDonnell had come to chat with Katzenbach's commission staff. James Vorenberg was out of the office, leaving Henry S. Ruth to act in his stead. Vorenberg came to the Justice Department at John F. Kennedy's behest, and President Johnson asked him to stay on to run the Commission's daily operations. Vorenberg would have wanted to know that McDonnell had stopped by.

Mr. Ruth was a hard-nosed attorney in his own right, someone with the temperament and tenacity to oust a president one day. He was meticulously organized. Precise. Someone who kept track of every word spoken. Ruth planned to brief Vorenberg fully on his conversation with McDonnell, which went something like this.[1]

What brings you here today, Mr. McDonnell?
    Well, Jerry Daunt suggested I stop by.
    Oh yes, Jerry, in the FBI's Uniform Crime Reporting Division. Great. Do you work in law enforcement?
    Not anymore. I used to though. Was a police officer out in California.
    Okay, and you have information you want to share with the Commission about your experience as a police officer?
    In a manner of speaking, yes. But I am here about something better. I'm with IBM now.

Oh, I see. That's a bit different.

Well, yes and no. I'm an account executive with our data processing division. I handle all of IBM's law enforcement work and products.

I didn't know there was such a thing. IBM products for law enforcement. Have you been in that position long?

I was promoted just a few months back. Right about the time the Commission was formed, in fact. That's what I want to share with you. I've been in Oakland, California, and Los Angeles, Chicago. ...We have rolled out a number of new products that I think will help law enforcement do their jobs better. More efficiently.

Their conversation was straightforward. Focused. This was a courtesy call for Ruth. McDonnell had interrupted his day. Ruth, though, was genuinely interested in what McDonnell had to say. But Ruth's goal really was just to get it all out on the record, and pass it along to Vorenberg and the Committeemen. Then be done. He nodded when McDonnell spoke. He did not necessarily agree. But, like a good attorney deposing a witness, Ruth's nods helped McDonnell keep talking. Not that McDonnell needed the help.

McDonnell made sure to let Mr. Ruth know that he wasn't new to federal law enforcement. He wanted to make sure it was clear to Ruth that IBM shared the government's interests.

We have five new or soon to be developed data processing applications that could be a great help to law enforcement. The first is a program that can store and pass on information. A second one processes fingerprints. Another can help departments deploy patrol and other personnel. The last one—and this is the one that is already in use and shows great promise—is an information management system that can help law enforcement officials. We call it a command and control system.

You mean the same kinds of systems the army and air force use in the field?

They would be adapted to law enforcement. But, essentially, yes.

Sounds interesting. Please continue. Are there law enforcement agencies that are using these already? Or you are testing them out?

Yes. For instance, we're already working with the FBI. Very likely we are going to help them develop a national database for crime information.

Interesting. And can you tell me a little bit more about why that's necessary? What problem could it help solve?

Well, such a system, with proper identification of goods, could be adapted to stolen autos, jewelry, appliances, office equipment, machine tools, guns, cigarettes, and liquor.

I see. So it could help keep track of stolen property all over the country and therefore help to locate both the property and perhaps the people who stole them?

Exactly.

And perhaps such a system could also keep track of other crimes—murders, rapes, armed robberies and such?

Yes. Absolutely. And the most important thing about that kind of system is that it can also provide a mechanism for sharing this information across law enforcement agencies throughout the country. Any agency or officer could instantly pull up this information when needed.

That sounds like it could provide some real advantages. What else?

Well, we're working on two other products right now that I think would really be of interest. By the way, I assume that our conversation here can remain confidential? I would appreciate if this information goes no further than the two of us. And of course Mr. Vorenberg, and the Attorney General.

Yes, of course.

We are developing a low-price, high-speed teletype which can be installed in every police car in an endeavor to counter the "airtime" problem which now exists with conventional communications facilities. I think they intend to experiment with this on the Chicago Police Department.

And when you say the "airtime problem"?

Yes. Speed. In essence, this would enable patrol officers to communicate more quickly with each other and with the central police command. But, there's another project that I'm really excited about.

By all means…

We are working on the possibilities of county-wide and state-wide computer storage of all information available in every government file concerning each individual. For example, we have started, on a pilot basis in California, to do this in an effort to counter welfare cheating, wherein a person may be on the welfare rolls of three or four different cities and also have a job in still another city. This development has obvious law enforcement applications.

Yes, I see where you're going with this.

McDonnell could have stayed there all day. He had much more to say. But he had accomplished what he had really come there to do: throw a little substance in between the winks and nods he sent Ruth's way. McDonnell wanted Ruth to know he had computers to sell. Software. Systems. Technicians. They were good partners willing to bend their technology to the government's needs—to its will. After all, software is more flexible, cheaper, and processes information more efficiently than a human being. As a former member of law enforcement, McDonnell made it clear. He placed his bet on the new systems IBM had started to build, and not in investing in more cops.

Ruth listened intently. Enough to document his conversation for Vorenberg. Enough to send McDonnell a letter two weeks later thanking him for the surprise visit. Ruth also asked McDonnell to send him any background materials about the systems and projects he had so intently discussed. As for McDonnell, he left the building as casually as he'd entered. He knew he'd be back. And even if he did not return, he knew his work was as good as done.

* * *

The Committeemen in that room at the Institute for Defense Analyses were the real deal. They were scientists, not salesmen. They knew systems and computers. Their methods were sound. Good enough to send men to space and back. Good enough to prevail over Axis powers. Good enough to make America the world's sole superpower. These were the ideas, methods, and experiments that they would bring to solving America's law enforcement problems.

The men were all aware that the task with which they had been presented also offered them an opportunity.

Science and technology is a valuable source of knowledge and techniques for combatting crime; the criminal justice system represents a vast area of challenging problems.

The Committeemen all seemed to have walked into the room agreeing with Katzenbach's statement. Apparently, none of them took issue with Katzenbach having described them as men with no background in law enforcement. They knew the power of science and technology exists sui generis.

The experience of science in the military, however, suggests that a fruitful col-laboration can be established between criminal justice officials on one hand and engineers, physicists, economists, and social and behavioral scientists on the other.

The Committeemen accepted this premise from Katzenbach too. But why? Why did the Committeemen assume that the military and domestic law enforcement were so alike? That they were similar enough to think that the technology transfer and the computational thinking that went with it would seamlessly translate from one field to the other?

Well, one commonality was clear. Both the military and law enforcement operated on a foundation of command and control. Simply put, command and control refers to strategic planning and coordination—the kind it takes to deploy people and tools to produce a purposeful, pre-designed outcome. Command and control thrives on information, communication, leadership, logistics, procedures, resource allocation, equipment, and technology. The marines—the nation's on-the-ground combat force—spent a lot of time thinking about command and control.

Command and control is *based on our common understanding of the nature of war.... It takes into account both the timeless features of war as we understand them and the implications of the ongoing information explosion that is a consequence of modern technology. Since war is fundamentally a clash between independent, hos-tile wills, our doctrine for command and control accounts for animate enemies actively interfering with our plans and actions to further their own aims. Since we recognize the turbulent nature of war, our doctrine provides for fast, flexible, and decisive action in a complex environment characterized by friction, uncertainty, flu-idity, and rapid change. Since we recognize that equipment is but a means to an end and not the end itself, our doctrine is independent of any particular technology.*[2]

But the Committeemen were realists. They didn't approach command and control problems in law enforcement like some Dirty Harry–style cops, someone who made decisions after a short conversation with his ample gut. In fact, one of the Committeemen pointed out that their role as scientists was to give the public scientifically generated options for how to reduce crime. Each would have its benefits. All would have—usually enormous— costs. But they felt it was up to the people—not themselves—to decide. It was not their role to determine what price they would be willing to pay to accept technology's payoffs.

Another member made it abundantly clear for the record. And he spoke for them all.

This is often a difficult decision to make, since for most inventions, no one can now say what they will do about crime—very little being known of what anything will do about crime. Inventions . . . make possible actions heretofore impractical. But their value in reducing crime is not known and will remain so until careful field evaluations are conducted.

The Committeemen thought about their ability to reduce crime in the same kind of time horizon that epidemiologists and pharmaceutical developers think about producing cures for diseases. Not next year. Not ten years. Maybe a quarter century if we get really lucky. Half a century more likely.

But the attorney general, President Johnson, a frightened white America, and IBM did not have that kind of time.

* * *

Tom Watson Jr. may not have had an official opinion about IBM's sponsoring CBS's Watts documentary. Who can say? But he did have an opinion on Watts. He voiced it to boost IBM's corporate responsibility narrative.

There had been race riots in the Watts section of Los Angeles. . . . Nobody wanted the violence to spread. As a liberal Democrat I feel a duty to solve such problems—but I don't know how.[3]

Watson had enlisted himself and the company in Lyndon Johnson's War on Poverty that Johnson initiated when he was elected in 1964. Watson searched for opportunities to be relevant. He said he wanted to help address the social ills that plagued society, particularly the plight of America's urban poor. But he was right. He didn't know what he was doing.

Through a subsidiary we were running a major Job Corps center at Camp Rodman, an abandoned army base in New Bedford, Massachusetts. The idea was to train seven hundred fifty "hardcore unemployed" each year—black high school dropouts from the inner city who had never held jobs. The experience caused us some real soul-searching, because there were more problems than we anticipated.[4]

Apparently, Watson's hardcore unemployed folks wandered away from their military ghetto and started roaming New Bedford's streets. Residents objected to their presence and their activity.

IBM ended up hiring very few Camp Rodman "graduates," and I doubt any other company did either.

The government eventually closed the project and gave up on it altogether. Watson and IBM continued to work on new ideas to help solve this problem. Meanwhile, his team of IBMers scattered throughout the Commission had already solved theirs.

* * *

The report was out. The Commission had labored for more than a year since they first gathered around that Institute for Defense Analyses conference table. Some of its work still continued, but its conclusions—sprawled over 342 pages—were expansive, crystal clear, and emphatic.

The scientific and technological revolution that has so radically changed most of American society during the past few decades has had surprisingly little impact upon the criminal justice system. The public officials responsible for establishing and administering the criminal law—the legislators, police, prosecutors, lawyers, judges, and corrections officials—have almost no communication with the scientific and technical community.

The Committeemen had worked as hard as they could to rectify this. And they made a good start, at least as far as the government and its private industry partners were concerned. Among their extensive research catalog, the Committeemen completed *The Los Angeles Study*. That project had analyzed 1,905 crimes reported to the Los Angeles Police Department in 1966. According to the Commission, the LAPD was *a notably well-trained and efficient police department*. The study's premise was simple. Police work is about solving crimes and catching criminals. Data from the study revealed that police solved 86 percent of crimes that had a known suspect. And, the faster police responded to a reported crime, the more offenders they caught and arrested.

The Committeemen reached a significant conclusion. Police would solve more crimes if they responded more swiftly to crime reports, or if they collected better evidence that led to credible suspects. To the Committeemen, this was a technical, not a manpower, issue.

Scientific crime detection, popular fiction to the contrary notwithstanding, at present is a limited tool. Single fingerprints can be used for positive identification when compared to those of a named suspect, but they are of limited utility when there are no suspects. There is no practical method for classifying and searching single latent fingerprints by a manual search of local, State, or national files.

This was one of many examples The Commission used to press for more and better law enforcement technology.[5]

The Committeemen also embarked on *The Washington Study*. That study had focused on the courts, rather than the police. The question? How to arrest, appear, and adjudicate suspects more swiftly and efficiently. The Committeemen's researchers entered Washington, DC, felony arrest data into a computer. Then they tested several modeled solutions to the problem. The computer delivered an automated recommendation.

It appeared that the addition of a second grand jury—which, with supporting personnel, would cost less than $50,000 a year—would result in a 25-percent reduction in the time required for the typical felony case to move from initial appearance to trial.

The conclusion seemed commonsensical. Did the Committeemen really need a computer to tell them that two grand juries could facilitate double the caseload? Nevertheless, the experiment served its purpose. It showcased the computer's data processing power. And, it demonstrated the computer's ability to model and automate complex decision-making. In this case, it made decisions about how to allocate human and financial resources.

The Commission made many recommendations based on the Committeemen's work. They all added up to a single conclusion: the federal government should spend hundreds of millions of dollars to build, test, and transfer new technological solutions to law enforcement agencies to help solve law enforcement problems. But what should these new systems focus on? Which of the many elements of law enforcement was the computer especially equipped to tackle? For what problems were law enforcement agencies seeking a solution? The answer to the final question was ... none.

The Committeemen launched their studies, gathered their data, crunched their numbers, plotted their systems, and redefined police operations. All the while, police chiefs and rank-and-file officers from around the country didn't give two shits about using technology to do a job they knew they had been doing so well all on their own.

And so the Commission, in order to justify its recommendations— established and described a whole list of potential use cases. Each was perfectly designed to deploy computer systems in law enforcement scenarios. The Committeemen's job, so it seemed, was to provide the evidence that government, and industry, could use to persuade, convince, or cajole police departments across the country that they should invest in the technology. The Committeemen gave them the ammunition they needed to coerce law enforcement to use computer systems to radically transform police work.

The Committeemen ended their work. The Commission filed its report and published its recommendations. And, it acted swiftly on them. Early in its work, the Commission had opened the Office of Law Enforcement Assistance. The new agency was the conduit through which the Commission undertook and commissioned the investigations it relied on for its report. Within a year of the Commission's conclusion, Congress passed more permanent legislation. It established the Law Enforcement Assistance Administration. From it would flow billions of dollars to state and local law enforcement agencies, all for the sole purpose of building what would come to be called *automated criminal justice information systems*.

For Katzenbach, the Commission, the Committeemen, Henry Ruth, and R. E. McDonnell—their work was done. But for others, it was just getting started.

Chapter Fourteen

WHAT HAPPENED AT
THE HOMESTEAD

The summer of 1967 was long and hot. No crime report could have staved off the impending revolution. Not even a report with such revolutionary recommendations as the Commission's. The revolution brought about by civil disorder descended on the nation's streets, from Detroit and Albina, to Chicago and Baltimore, Englewood and Flint, Grand Rapids and Houston, Rochester, Summit, Tucson, and Newark. There were many, many more. Watts repeated itself, and for the same reasons. Police brutality. Unchecked surveillance. Disproportionate criminalization. Exclusion from schools, education, job training, and jobs. Not to mention the utter lack of humanity with which America responded to black America's plight.

The uprisings took their toll on black communities and the nation. They stretched police capacity and raised the specter of domestic militarization. They produced carnage, death, and destruction to property and infrastructure into the billions of dollars.

The revolution was here. But the American legal system called it a civil disorder, mass civil disobedience. But unlike bus boycotts and sit-ins, civil disorders openly, vociferously, and violently defy those in power. Rioters— that's what they always call those who obstruct authority—obstruct the law and law enforcement agents. Civil disorders are revolutionary because they threaten the established social, political, and economic order. At the very least, they stare power in the face and force it to flinch.

When Johnson rolled out Executive Order 11365 at the end of July 1967, he named the newly established investigative body The National Advisory Commission on Civil Disorders. Its name was a sign that the nation started to feel black people's revolutionary threat. Johnson chose Illinois governor Otto Kerner to head the group. Johnson appointed the kinds of usual suspects—judges, senators, lawyers, and such. NAACP head Roy Wilkins was the only black member of the group. The advisory commission's task was simple, focused. How did these disorders begin? How do we control them? How do we end them—forever?

* * *

Poughkeepsie, New York, is a small Hudson Valley town. It is home to just under forty thousand residents. It is quiet and secluded—a place where New York City folks who thrived on the hustle and bustle come to summer, sometimes winter, and often to retire. Tucked away in the middle of town sits a sprawling and stately brick mansion. In fact, it had been a Masonic lodge, until IBM purchased the property in 1939. It served as IBM's headquarters for business diplomacy and housed foreign dignitaries, heads of state, and other VIPs. They named it The Homestead.

On November 9, 1967, Kerner had summoned his commissioners to Washington, DC, for one of the Commission's regularly scheduled meetings. They gathered for a full day of hearings in the Executive Office Building, room 446.[1] There, they heard a string of white academics wax eloquent about the black family in the ghetto. The most prominent among them was Eliot Lebow. Lebow had authored an acclaimed book that year called *Talley's Corner: A Study of Negro Streetcorner Men*. The book charted the social relations developed among a group of unemployed black men who gathered every day at the same street corner in Northeast Washington, DC. Daniel Patrick Moynihan had called the book a work of brilliance.

At noon, the commissioners ate an informal lunch. Then they digested a briefing given by members of the Commission's Advisory Panel on *Insurance in Riot-Affected Areas*. For the afternoon, they engaged another round of hearings about the administration of justice during civil disorders. The Commissioners were left to themselves for the evening. Kerner hoped they would take advantage of their free time and pore over the mountain of documents to prepare for the important business meeting scheduled the following morning.

The next morning, the delegation gathered in a meeting room at the Capitol Hilton, a luxury hotel just north of the White House. No doubt they slept comfortably through the night. Hotel staff fully accustomed to satisfying the whims of the rich, famous, and powerful catered to their every need.

At the morning meeting, members received a briefing that prepared them for the upcoming weekend. They retired for lunch and then got ready to take their leave. Promptly at 2:50 that afternoon, the commissioners loaded onto a chartered bus and began their sixteen-mile journey to Andrews Air Force Base. The gentlemen boarded their plane and flew for ninety minutes. Then the plane landed at a municipal airport in Poughkeepsie.

Power. Politics. Diplomacy. Deal making. That stuff was Washington's daily routine. So what compelled Kerner and his commissioners to leave their normal political habitat to travel to a secluded dwelling three hundred miles away? What could they accomplish there that they could not at the posh Capitol Hilton or in the Executive Building? That was yet to be seen. For the time being, the gentlemen settled into their rooms at the Homestead. Limousines chauffeured them over to the Poughkeepsie Golf and Country Club for a brief reception. They returned to the Homestead for dinner, then retired for the evening. As the commissioners slept, their clandestine agenda waited patiently to greet them the following day.

* * *

The meeting attendees called it a "media conference." It felt more like an intelligence briefing or a special operations strategic planning session for the counterinsurgency. Their cover was tight. America's riotous summers took place within a novel news media environment. They weren't sure just how, but President Johnson was quite certain that television news especially, and newspapers as well, had fanned the flames of racial discontent.

Civil rights leaders and rank-and-file Negro citizens viewed the media as discriminatory. As far as they were concerned, the media only reported news in the ghetto when violence occurred or when crimes that affected white people took place. The media fed white people's stereotypes. They thought Negroes everywhere were nothing but a bunch of violence-prone thieves, thugs, and miscreants. Not that white people needed much help believing that. But from the standpoint of the government and law enforcement, the news media, at worst, sensationalized the riots. This, they speculated, just invited Negroes to give in to their deviant impulses.

The motive behind the Homestead "media conference" was simple and in some ways naive. But it was certainly not counterproductive for those attuned to its hidden agenda.

First, it will enable representatives of the media from all parts of the country and different types of cities and publications to get together in a relaxed, informal setting where they can discuss the inter-relationship of the press and race relations, and perhaps begin to articulate some of their concerns and their approaches to the problems in this area. Second, it will permit media representatives to meet with members of the Commission in an atmosphere, hopefully free of the suspicion and hostility that might otherwise surround this kind of inquiry.[2]

This was a perfect lure. Journalists gained what they crave most: access. The politicians and government types got an opportunity to apply some gentle persuasion. No doubt some of the media professionals invited sensed some ulterior motive. But they also knew that Washington powerbrokers could do little to shape news practices. Regardless, there was something in the program for everyone. And, enough to keep anyone from being suspicious.

The heads of the three television outlets—CBS, NBC, and ABC—were invited, as were key editors for the *Washington Post, LA Times, Newsweek,* and other such outlets. They had both national standing and local relevance in areas where riots broke out. They added a couple of token Negroes to the list. This who's who cast of American news media and journalism may not have been apprehensive about the Kerner Commission's motives. But that did not mean there was no reason to be suspicious.

Who had decided which long list of journalists to invite in the first place? Harvard Law professor Abram Chayes had assumed this dubious task. The former John F. Kennedy aide signed the memo, but not even he really called the shots. Nor, it seemed, did Governor Kerner. Or, Burke Marshall for that matter.

Rather, the fingerprints of the Simulmatics Corporation—an experimental corporation of dubious origins—were all over the weekend's agenda.

Saturday's theme was *the media in the riot city.* The first event began at 9:30 A.M. It was a "Television Roundtable" lead by Burke Marshall. Marshall was a crucial part of the event's thinly veiled cover. He headed the Department of Justice's Civil Rights Division when Robert Kennedy was the Attorney

General. Nevertheless, the interests of the men who orchestrated the panel lay bare and naked in the program's description.

The morning session's objective focused on the *problems of liaison and communication between* media personnel *and* police, city government, *and* ghetto leaders. This included:

> 1) How to establish effective lines of communication between television editors and official police sources; how to make these channels operative when a disturbance begins; and how to keep in touch with the activities and ideas of city officials; 2) Problems of communication for reporters covering riots: *getting to the police;* checking out *official information;* communicating the situation to editors; and 3) The problem of rumors: methods of checking reports of disturbances and verifying information—(a) *within the ghetto;* and (b) with *police and city officials.*

The same line of questioning was repeated in a concurrent session for newspaper editors and journalists. Should there be controls? Should television producers or news editors report whatever they saw or heard, just because they saw or heard it? Should information be verified? If so, by whom and through what channels? Burke Marshall and Dick Baker (an associate dean at Columbia University) led their respective media representatives in conversation.

The day's afternoon themed session—*the media in action during a riot: deploying the forces*—spotlighted a similar interest pattern. Kerner's Commissioners weren't that interested in learning from the newsmen they had assembled at the Homestead. They primarily wanted to instruct them about how to best cover a riot. Even more, they wanted to control the information disseminated by the press during a riot. Their admonitions to the journalists in the room boiled down to this. First, report information from official sources. This meant information that came directly from law enforcement and other authorized public officials. Second, at the very least, do not obstruct—by way of equipment or information—law enforcement's work. And, if at all possible, report in ways that facilitate that work.

Sunday's agenda was much the same. It promoted similar law enforcement and government interests. But the motives for Sunday's program altered slightly. The goal was to cast suspicion on any information that came out of the ghetto. Political leaders feared that television news cameras directed at certain Negroes would amplify undesirable black voices, those of other ghetto "agitators." They were careful not to create revolutionaries by puffing up some Negro hatemonger's ego. Black Power, from their point of view, needed to remain a slogan, not a reality for any individual or group.

Again many masked their true interests by asking questions that seemed to have Negroes' best interests at heart. Should newspapers cover more mundane, everyday ghetto goings on instead of just reporting the latest murder or riot? How do you get stories about what is going well in the ghetto out to the white masses? And, how do you get white people to listen?

The Homestead "media conference" was a facade. Sure, it provided an opportunity for media to consult men of renown and power. But the caliber of folks invited? The media already had access to these kinds of state power-brokers. Yes, it was lavish to be treated like the kind of stately royalty that typically took residence at the Homestead. But frankly, these men did not need the trip. Kerner's commissioners knew there was more to this trip than what met those newsmen's eyes. They knew the weekend was about something more, something bigger, something of greater consequence that would have real impact. They knew that they were all merely frontmen for the weekend's real agenda—an agenda purposefully designed in its entirety by the Simulmatics Corporation.

* * *

In the early 1950s, MIT decided to infuse its science and engineering curriculum with the social sciences—political science, specifically. But political science would be no island. MIT expected that its new professors and research programs would conform to, not deviate from, the institute's mission. Political knowledge would be practical knowledge. It would nurture and support the institute's engineering and scientific work. The institution would direct insights into social organization, social processes, and political power toward their industrial, corporate, and political interests. It would help solidify the nation's political and economic power. And when necessary, it would champion the interests of select persons and groups along the way.

In its first political science professor, Ithiel De Sola Pool, MIT found its champion. His first effort to fulfill the institute's vision would set its sights on nothing less than the heart of American political power: the presidency. Pool was a student of public opinion. He believed that human social systems and behavior could be simulated. Given enough of the right data, Pool and his associates knew they could conceptually mimic how people make voting decisions. If they could conceptualize this process for a single individual, they could do the same for one million, collectively. They believed they could do so by treating them not as individuals, but as members of particular groups with which they strongly identified—that is, their political party, religion, social class, or race. These all connected people together in the United States. Race, especially. Once replicated, Pool could represent the process as a mathematical equation. Scientists could then manipulate and optimize the algorithm—and perhaps people's propensity to vote for a particular candidate. If only Pool could perform this kind of scientific magic!

Pool claimed he could do just that. And he was ready to make believers out of the scientific and political community. He announced that he and his associates could not only predict, but influence the outcome of the 1960 US presidential election. Democrat John F. Kennedy faced Republican Richard Nixon. What would Pool need to wield this kind of scientific and political power? And if Pool could do this in 1959, why had no one tried before him?

Pool had a simple answer. No one had access to MIT like he did in 1959. More specifically, no one like Pool, with his interests, had ever before had access to MIT's Computation Center, which had just opened its doors two years before.

Pool said his scientific alchemy required four elements. He needed a massive amount of data. Specifically, he needed polling surveys about how people had intended to vote in prior presidential elections. He also needed these surveys to contain *standard identification data on region, city size, sex, race, socioeconomic status, party and religion.* Pool was keenly aware that Americans—white and black—were growing more concerned about race issues. He also knew that political elites knew very little about how Kennedy's Catholicism would influence voters' decisions. Ultimately, Pool sensed that race and religion would have the greatest impact on the 1960 election.

Pool found his massive store of data a little more than a hundred miles west of Cambridge, Massachusetts, in Williamstown. There, Elmo Roper's Public Opinion Research Center had been collecting data from public opinion surveys since 1939. This included polls from 1952, 1954, 1956, and 1958. It included data from one hundred thousand surveys. Pool called this his data bank, a place from which *one might draw the answer to any one of a vast number of questions.*

But Pool still needed an analytical process suitable for manipulating this massive data store. Using computational procedures, Pool and his colleagues aggregated the survey data. From it they developed 480 voter profiles. These profiles represented the range of socioeconomic categories, *Eastern, Metropolitan, lower income, white Catholic, female Democrats,* for instance. Next Pool correlated each voter profile with fifty issue clusters. What resulted was a matrix that plotted voter profiles with their positions on political issues. From this Pool could develop his simulation. The computerized simulation would run through every voter profile/issue cluster scenario and determine how each would likely vote. Pool would then know how likely, for instance, urban, Protestant, female, middle-income Negroes would be to vote for Kennedy. Kennedy could be advised to plan accordingly.

One of the reasons that no one had attempted this before Pool was simple. The necessary processing tools had never been available. That changed with the advent of the electronic computer. Poole explained.

Computers *can be used to follow step-by-step the logical consequences of series of events occurring in systems so complex that no formula exists for reaching the desired results. In such systems, which can be described but are not capable of being mathematically optimized, simulation permitting us to state the initial conditions and explore the consequences of specific changes over time. Simulation is, in short, an acting-out in the computer, of a history of events within a system. It replicates step-by-step the processes as they occur in time.*[3]

The IBM 704 housed at the MIT Computation Center provided the processing power Pool needed.

Pool's conceptual, analytical, and computational methods reflected the new field of operations research. And it didn't take long for Democratic Party operatives to get wind of its purported powers. Pool and his colleagues laid out their proposal to these men. Their primary objective, of course, was to snatch the presidency from the Republican Party. To pursue their interests, these party operatives formed the Simulmatics Corporation.

Mathematical simulation reflected operations research's essence: its principles and its application. Pool gathered together an advisory board. That team included the world's most renowned public opinion researchers and statisticians—Harold Laswell, Paul Lazarsfeld, Morris Janowitz, and John Tukey. In consultation with them all, the Simulmatics project moved from proposal to business plan. Having only the days between nomination and election day to work, Simulmatics produced successful results.

Simulmatic's researchers not only predicted Kennedy's success. In many ways, they manufactured it. Among its scientifically derived wisdom, Simulmatics had advised Kennedy to push the civil rights issue. Not necessarily because it was morally right. Not because it was the best for America's democracy. The advice to make civil rights a Kennedy campaign centerpiece was a scientific response to a political predicament. The Democratic Party had a Negro problem. Particularly, it had been hemorrhaging them in droves over the prior two presidential election cycles.

* * *

By 1967, Simulmatics had inserted itself into America's military industrial complex. On October 12 that year, Simulmatics reached out to close the loop in a conversation with Mr. Arnold Sagalyn. Sagalyn represented Kerner's Commission.

Dear Mr. Sagalyn:

> In accordance with your request during our telephone conversation of October 12th, 1967 we are sending you a copy of our report entitled "Urban Insurgency Studies." I have requested Col. Yates, who runs our Cambridge office to send you a copy of the write-up on our Community Game.

Dr. Jules David Yates, representing Simulmatics, referred to its Community Interaction Game. The live, interactive game relied on humans to assume similar profiles as the election simulation. In this case, however, human beings role-played their decision-making process based on the scenarios they were given.

Soon after Yates's correspondence with Mr. Sagalyn, the Kerner Commission contracted with Simulmatics to undertake a "media study," which Simulmatics coordinated.

Simulmatics initiated a research project guided by a straightforward and precisely circumscribed mandate. The basic question to be answered was: "How did television newscasts and newspaper reports in a selected number of American cities present news coverage of the 1967 summer riots and the race relations background of these disorders?" As a secondary inquiry Simulmatics sought to examine the question: "What reaction did various audiences have to the media and what effect, if any, did these audiences believe the news media had on the riots?"[4]

Kerner's Commission had wrestled with a systemic problem. How do we curtail the revolutionary threat to America's racial, political, and economic order? Answering this question required the kinds of behavioral and demographic data that Poole had used in his 1960 presidential election simulation. But Simulmatics would not be concerned with America's election system for this project. They needed to understand the kinds of urban riots that had played out in recent months and years and map its contours. They needed to understand how these potential revolutions materialize. Who leads them? What sparks someone's interest in participating? How does information about them spread among both would-be revolutionaries and those they threaten?

Simulmatics already had a theory. To them, riots and revolutions were, more than anything else, a specialized propaganda war. What better way to understand the information and communication infrastructure that powered urban riots than to conduct a detailed, large-scale operations study? And what better way to hide one's national urban intelligence objectives than by masking them with a high-powered media conference, and conducting that conference in secret at The Homestead?

Kerner's Commission wasn't Simulmatics' only government client at the time. As it entertained its new contract, Simulmatics was dotting its i's and crossing its t's for a final report. A report for Advanced Research Projects Agency Order number 877. Simulmatic's work was monitored by the Army Research Officer under Contract No. DA 49-092-ARO-152. Code name: Project AGILE.

Simulmatics' subcontract with Project AGILE pursued a singular goal: to improve the effectiveness of the Chieu Hoi program in Vietnam. Chieu Hoi was a propaganda and psychological warfare campaign. The US mobilized it to coerce Vietcong insurgents to defect from the National Liberation Front to side with the government during the Vietnam War. Simulmatics' media riot study made use of the same underlying methods and tactics it had used to develop Chieu Hoi.

The study was conducted by three-member teams chosen by Simulmatics for their knowledge of ghetto communications, city life, and experience in dealing with and understanding social problems. Each team member, individually or as a group, conducted both short and in-depth interviews in the ghetto and downtown areas. Information was developed in the course of brief encounters with ghetto residents of the riot city, residents from the adjacent riot areas, and non-ghetto residents of the city. Interviews were conducted on public transportation vehicles, in restaurants, taxi cabs, pool halls, grocery stores, dance halls, street corners, barber shops, subway, bus and train terminals, hotel lobbies and various stores and businesses located in the main downtown area. Lengthy and in-depth interviews were also conducted in homes of residents from the riot area, and on occasion in churches of the riot areas. Several professionals, such as doctors, lawyers and community leaders from the riot areas were interviewed at length.[5]

Simulmatics was interested in understanding how Negroes felt about the media's coverage of the riots. But they were also interested in the *movement of people in and out of the city; movement of people in and out of the riot area; anticipation and/or preparation for and during the disturbances; disruption caused by the disturbances.* To decipher this they investigated everything you could think of: traffic reports to toll booths and other junctions leading into and away from riot cities, gasoline sales, bus traffic, and more.

Simulmatics had to determine who the enemy was, identify its leadership, and graph the channels they used to distribute information. Then the job was to determine what motivated this group to engage in anti-government activity and persuade them toward pro-government activity—and if unsuccessful, to neutralize the activity. The Kerner Commission's would-be puppet masters figured they could leverage Simulmatics' success in Vietnam by applying the same tactics throughout America's inner-city corridors, in Los Angeles, Detroit, Newark, Harlem, Atlanta, Tampa, New Haven, Milwaukee—the whole of black America. After all, the government fought the same enemy in both places. It was revolution time.

* * *

What white Americans have never fully understood—but what the Negro can never forget—is that white society is deeply implicated in the ghetto. White institutions created it, white institutions maintain it, and white society condones it. Our Nation is moving toward two societies, one black, one white—separate and unequal. Reaction to last summer's disorders has quickened the movement and deepened the division. Discrimination and segregation have long permeated much of American life; they now threaten the future of every American.

These two statements summarize the report that Kerner's Commission delivered to President Johnson. The body's findings—and even more, their prescriptions—simultaneously shocked and appalled the president. Turns out it was his prerogatives that led to the Simulmatics project. Johnson was the one deeply skeptical about the urban uprisings. It turns out he was the one who believed them to be insidious and communist inspired. Also, Johnson wove a web of conspiracy in his own mind. His fantasies implicated and drew the American news media into a conspiratorial triangle that included the Negro and the Reds. Kerner's Commission rejected it all. And Johnson rejected Kerner's report. He was better off fighting the Vietcong, he thought. So that's where he invested the nation's resources.

The Kerner Report detailed the tragic outcome that white racism wrought on black America. It also showed the nation a way forward. But the nation brushed it aside. The greater tragedy, however, was that Washington, and the emerging computing research and development establishment, tipped their hand. They would increasingly harness the computers' hardware and software capabilities to further oppress, rather than liberate, the Negro.

Simulmatics had been wildly successful, even though its results were ultimately buried alongside the commission's report. Using Simulmatics for this work, however, legitimized and normalized the principles on which it was based: the idea that the computer could model, and therefore manipulate, human systems and behavior. It once was theory. It soon became policy. Black people would continue to remain its subject for experimentation. Computing power would be used *on* them.

Chapter Fifteen

KANSAS CITY BURNING

Kansas City is the country's geographical center. By 1968, the Great Migration drove close to 120,000 blacks to it. They made up 20 percent of the city's population. Most were densely concentrated into the city's urban core. They called it "East Kansas City."

Troost Avenue stretched from the Missouri River on the city's northern tip down to Indian Creek at its southern. It marked Kansas City's color line. In 1968, the *New York Times* described the city's segregated, black core.

While the East Side is a ghetto, it is not a teeming slum. Much of the area is made up of tree-shaded streets lined with good, one-family homes.[1]

That's how they described Watts, just before the 1964 uprising. And Kansas City—like Los Angeles before it—felt immune from racial unrest.

While Kansas City ranked only twenty-sixth among the United States' most populated cities, the municipality boasted a big-city police force. It secured a chief to match. A Kansas City native, lawyer, US Navy veteran, retired FBI agent and supervisor, Chief Clarence Kelley represented law and order not just for Kansas City but also for the nation. Kelley was particularly known for his expertise in containing riots and other outbreaks of large-scale, violent, civic resistance.

A reporter once described Kelley as a *squarely built veteran of 22 years with the FBI,* and *a veteran lawman, a professional lawman who talks with confident ease about the techniques of containing insurrection.*[2] Kelley, the lawyer, recognized that civil disobedience properly exercised a citizens' right to petition the government for redress of grievances. And he believed such demonstrations were capable of being peaceful and nonviolent, and that they could thwart, rather than stimulate, criminal activity.

However, Kelley, the cop, believed that *it is difficult for poorly educated and unsophisticated groups to make the fine distinctions between a demonstration and a riot situation. The type of disobedience now being encountered . . . mass violence and force such as utilized in Watts, Harlem, Chicago . . . create a state of incipient revolution.*[3]

The urban "riots" of 1967 had stoked Midwesterners' (and other Americans') fear of crime. The more they voiced their fears, the more Kelley spoke out. He exclaimed the need to properly equip law enforcement agencies to minimize threats to law and order.

Obviously, we just have to be much better at our work. We need all kinds of new equipment, scientific equipment of the highest order to help us out. We could prevent a lot of crimes if we had the gear and know-how that is available to a lot of the rest of society.[4]

President Lyndon Johnson's Crime Commission had shared Kelley's sentiments. They had agreed with both the justifiable and pervasive fear of crime. They had agreed that the nation must better arm law enforcement agencies to meet the looming violent threat. *THERE IS MUCH CRIME in America,* the Commission's Report, titled "The Challenge of Crime in a Free Society," began. *Every American knows that.*

Whether Kansas City's public officials or its citizens engaged the report's findings or not, one thing is clear. They were not prepared to experience them first-hand.

* * *

Black youth personified and represented the threat pulsating through the nation's urban ghettos. Their presence stoked the pervasive fear that ghetto problems could invade white America's protected sanctuaries. After all, the only things that really separated and protected them were flimsy structures built into the geographical environment: train tracks that separated black communities from white, long streets that demarcated racial territory and psychologically policed black mobility, highways that were designed and erected to deter the free and easy movement of the criminal element into the city's economically thriving urban zones, and suburban white ethnic enclaves. But none of these measures could compete with tear gas, fire, or people with nothing to lose.

However, there was no threat that April 9, 1968. That day, the only thing Kansas City's black youth wanted to do was take time to remember, reflect on, celebrate, and honor the life of their fallen hero, leader, prophet, role model, father figure, and hope: Martin Luther King Jr. Many cities—Washington, DC, Chicago, Cincinnati, Pittsburgh, and New York City, to name just a few—erupted the night of, and following King's assassination on April 4. But all remained relatively quiet in East Kansas City. Many of its neighboring schools across the border in Kansas, elsewhere in Missouri, and around the nation canceled school on April 9, the day of King's funeral. Kansas City schools, however, remained open.

The city's black youth protested at Central Junior High and Central High School. They staged a walkout and marched quietly to the mayor's office. Some wore signs with slogans like "We Shall Overcome." There they met an intractable mayor and a recalcitrant police chief. A thin, blue line of police officers and highway patrolmen stretched end to end in a semicircle in front of city hall, armed with rifles and donning gas masks, standing at attention. It was as if they protected the literal gates to the city from being stormed by a foreign enemy.

Meanwhile, a belligerent black woman, elegantly dressed with an attractive headscarf, skirt, and short-sleeved blouse, carried a handbag on one arm. She walked down that police line hurling epithets at the officers as she pointed toward them. She stood mere feet away from them, but she looked like she wanted them to clearly hear her anger. They would answer to her judgment. One officer told what happened next.

All of a sudden I saw a soft drink bottle from behind her, come flying over her, and broke. The next thing I heard was "click," "click," "click." That's the sound of a pin bein' pulled from a gas grenade.[5]

Police Chief Kelley had authorized the use of force. From the time that bottle hit the ground, the police waged war on the community. The mayor and Kelley called in National Guard troops. All evidence—at the time and in hindsight—shows that the police overreacted. Yet, as soon as the first pin popped, Kansas City had become just another tale of urban, black thuggery and lawlessness.

Mayor Davis marshaled the same riffraff theory peddled in the aftermath of Watts and other uprisings. He claimed that the majority of those eventually arrested were either young people, or people with criminal records. The mayor insisted that *it is apparent that the criminal element was using Tuesday's unrest as a smokescreen to attempt to cover their usual activities.*[6]

The mayor's words revealed what everyone—black and white—seemed to know. Consciously or subconsciously, white people believed blacks bent toward criminality. Their concentration on the city's east side was a containment strategy.

Emmanuel Cleaver—a Negro who would one day become East Kansas City's congressman—said, *31st Street, which had a lot of African American businesses, was essentially burned to the ground.*

The entire city skyline is lit up with flames. The worst seemed to be from about 27th to 39th Street to around prospect. Every one of those blocks is a war zone.[7]

People frequently asked—then and now—why black people would protest injustice by destroying "their own" neighborhood. Cleaver offered a simple explanation: *Because it was the only world they knew. To go West of Troost was to die.*[8]

The riots, the riffraff, the race—all stoked and confirmed white Kansas City's worst fears. The geography of race and crime among Kansas City's white community cried out for more, better, and more efficient protection. With the help of the federal government, Kansas City police chief Clarence Kelley would soon deliver.

* * *

The President's Crime Commission report in 1968 had recommended that the federal government invest massive amounts of resources into what were later dubbed Criminal Justice Information Systems. It invested millions of dollars to design and build them. The growing and persisting fear of crime was its underlying rationale. But the commission's long list of use cases for these systems ultimately proved most persuasive.

The computing industry, led by IBM, the federal government, national and local law enforcement agencies, and academics at elite science and engineering institutions had started developing these use cases beginning in 1965. That's when New York City police commissioner Harold Leary formed the *Joint Study Group.*[9] This study group included representatives from the police department's planning and communications departments and four representatives from IBM. One was a sales manager. The other three were computer programmers.[10]

In the end, the Joint Study Group outlined thirteen potential new law enforcement computer applications. The list included applications for computer-aided dispatch, crime analysis, fingerprint identification, resource allocation, and election returns.[11] New York City began pursuing only one of these identified systems—a computer-aided dispatch system. They called it SPRINT—Specialty Police Radio Inquiry Network. The system was built from scratch. But builders based it on an existing IBM design model for a flight reservation system.

During the same time, Kansas City's chief of police Kelley had assembled a team of his own. It consisted of an in-house team of two: his assistant, Lt. Col. James Newman, the department's chief data systems director, and Melvin Bockelman. Both were dubbed "patrolmen programmers." They were policemen first, but they were armed with technical data processing training. Two IBM personnel, marketing representative Owen Craig and Roger Eggerling, an IBM systems engineer, rounded out Kelley's team.

IBM described its systems engineers as *assisting our customers in defining their systems problems and determining the best combination of IBM equipment to solve them*.[12] Speaking more holistically about how IBM built its enterprise, the company had reported to its board and shareholders back in 1961 that *this era demands a higher degree of professionalism than ever before among the sales representatives who initiate and develop customer interest, the systems engineers who help our customers study, define and develop solutions for their problems, and the customer engineers who install and maintain equipment at peak efficiency*.[13]

IBM systems engineers were also its link to the scientific and engineering academic community. They presented seventy papers in one year alone, for example. IBM systems engineers refined their computing knowledge within an academic field. They had also distributed their knowledge about systems building throughout both the scientific and industrial community. The plan?

Imbed the police beat algorithm within a geographical crime information system with graphical inputs and outputs, thus enabling us to bring the proper man-machine interaction to bear on this heuristic-analytic type of decision problem.[14]

IBM systems engineer Saul Gass worked in IBM's government services division. Gass divided command and control systems into the two primary problem areas they confronted: police planning and police operations. Police planning had much to do with allocating human and material resources. How many police personnel should be dedicated to a given geographical area based on its population size and crime rate? How should you divide up a geographical area into efficient police patrol beats? How much equipment should be stored, and in what locations, in order to be ready to respond swiftly and effectively to a riot situation? These are examples of planning problems that police had to solve in order to maximize success.

Operational problems, on the other hand, involved different types of questions. How do you identify crime patterns? How do you both predict and apprehend suspects based on those patterns? Once apprehended, how do you associate suspects with other crimes they may have committed? And, when you know all this, how can you prevent crime from being committed in the first place?

These concerns were packaged into a command and control solution called computer-aided dispatch (CAD). Underlying the CAD system was software, powered by an algorithm that automated solutions to specific operational and planning problems. Its task was to answer the question of how to allocate a finite number of police patrol units to police beats (parsed geographical areas). And, how to allocate those resources to patrol beats so that police officers were positioned to be dispatched to and arrive at the scene of a crime. Gass's mathematical model could be used to determine this, given some known factors and data. He had already developed such a model. He also possessed "real-world" crime data, from New York City's SPRINT.[15] The array of symbols, functions, and notations looks complicated to the non-mathematician, but the information and data the algorithm called for tell us everything we need to know.

First, US Census tracts parse geographic areas and develop uniform, structured data about those areas—primarily population size and racial demographics. These tracts enable strategic deployment of police officers by geography, population size, and racial composition.

Gass's model (and the police community) contended that all crimes were not created equal. Thus, Gass's algorithm required "weighted" crimes. Like census tracts data, a police department like Kansas City's could rely on an existing weighting system. In the mid-1960s, the International Association of Chiefs of Police had already produced such a ranking.[16] A score of four represented the highest-priority crime. A score of one was the least threat, and therefore least priority. Criminal homicide, forcible rape, robbery, aggravated assault, burglary, larceny, and auto theft all received a score of

Minimize

$$\sum_{i=1}^{n} \sum_{j=1}^{n} (d_{ij}^2 c_j)\, x_{ij}$$

subject to

$$\sum_{i=1}^{n} x_{ij} = 1 \qquad j = 1,2,\dots,n$$

$$\sum_{i=1}^{n} x_{ii} = k$$

$$\sum_{j=1}^{n} c_j x_{ij} \geq \frac{aC}{k} x_{ii} \qquad i = 1,2,\dots,n$$

$$\sum_{j=1}^{n} c_j x_{ij} \leq \frac{bC}{k} x_{ii} \qquad i = 1,2,\dots,n$$

$$x_{ij} = 0 \ or \ 1,$$

where

k = *number of beats to be assigned*

n = *number of census tracts in city*

x_{ij} = $\begin{cases} 1 \ if \ tract \ T_j \ is \ assigned \ to \ the \ beat \\ \quad centered \ about \ tract \ T_i \\ 0 \ otherwise \end{cases}$

c_j = *the weighted crime workload in T_j, e.g. if I_{pj} = level of crime incident p in T_j then $c_j = \sum_p w_p I_{pj}$, where w_p is the weight of the pth incident.*

Figure 15.1. Algorithm developed by Saul I. Gass to divide police districts into patrol beats.

four. These were also known as "index crimes." The FBI had developed this system for its Uniform Crime Reports.

In addition to weighted crimes, Gass's formula required weighted crime incidents. And it required weighted workloads. Then, the algorithm required that police correlate workloads with the geographical areas where the greatest numbers of the highest-weighted crimes took place.

Using census tract designations, and these crime weights, Gass's formula *used five measures of the workload for a census tract: number of index crimes, population, area, level of crime multiplied by the population, and the level of crime multiplied by the area.* This produced a geographical map of a city, parsed by patrol beats. They could be designated as high to low threat. These criteria could then be used to determine police resource allocations. One might, for example, assign twelve police officers to regularly patrol the high-threat area, and only three for the low.

It could also be used to determine whom and how many police officers to dispatch to a given area when a crime was reported. It would determine with what urgency and speed the officer(s) should respond. And it determined what precautions police should take in order to protect their safety. A call reporting a "suspicious" Negro loitering in a low-threat area, for instance, might lead a dispatcher to hail four squad cars. The Negro profiled as high threat; the neighborhood coded as low threat and white. Of course, one need only correlate these threat areas with their corresponding census tract demographics to begin to formulate not only geographically based threat profiles, but the corresponding racial profiles as well.

Producing and then systematizing such a profile in ways that could have measurable effects, however, required a much larger system. It would have to include more applications than just CAD. It would need to be networked; reach beyond a single city or local area; and be able to constantly ingest new data, process that data, and use them to model criminal profiles and affect future police decision-making. Such a system would be a massive undertaking. It would cost millions of dollars. Those who commanded it would be compelled to demonstrate that the system's outputs produced the desired outcome: to efficiently protect America's white citizens from its most feared criminal suspects.

Chapter Sixteen

THE MAN'S BEST FRIEND

The magazine headline called it "A Cop's Best Friend." The article that followed conjured an idyllic scene. Small-town life nestled within a big city's infrastructure. Cops patrol city beats. Citizens call for assistance. Police officers swiftly deliver safety and security with the aid of a computer stationed at police headquarters. The need for police protection at all is the city's most unfortunate byproduct. But those in command have everything under control. That was the magazine's story. But the photograph that frames the story shows that there is much more behind the curtain.

There Chief Clarence Kelley stood, authoritative, inside Kansas City's police headquarters. To his right, two large screens displayed city maps, divided into varying shapes and sizes. Icons mark key data about each section. The map is under control, which means the city is under control, well managed, taken care of.

In the same scene, Kelley locks eyes with a single police officer, seated at a computer terminal. Headset affixed, the computer operator keystrokes data into a computer. The computer, in turn, delivers information to a telephone operator. The operator delivers the final relay to a patrolman out on the beat. The scene speaks nothing of the fires that raged and torched Kansas City's east side just four short months before. But make no mistake. Kansas City remembered the fire that burned that time, and feared the one that could come next. That was why the new computerized command and control system existed.

* * *

Kansas City's ALERT II system debuted in August 1968. Parts of the system had been tested starting back in 1964. That's when Kelley first gathered together the necessary expertise to build a law enforcement information system. And in just four short years ALERT II put into place almost every law enforcement use case the Committeemen had once previously proposed to the President's Crime Commission.

ALERT II required a detailed owner's manual. It described the system, which seemed harmless enough, routine, and natural. And that was the message the system's designers and Kelley wanted to convey. The new computer system would become part of the police department's family. One of their own. A benign tool to help good police officers do what they do better, which is fight crime and process large amounts of routine paperwork.

But that manual was as carefully designed as the system it described. Don't get me wrong. A lot of sinister shit lay buried between lines and lines of technical jargon. But before readers got to it, ALERT II's story needed to be told. On the manual's front cover, a police vehicle sits somewhere in the field, scenes from Kansas City's skyline set off in the distance. An officer rests in the front seat of the car. His head bows toward the car's steering wheel, patiently awaiting that call from the dispatcher. She will tell him where they need him and his vehicle next.

Awkwardly positioned as if sitting in the squad car's back seat, another officer appears seated at attention. A massive, L-shaped IBM computer console surrounds him. A grid-lined map hovers just above. An array of input, output, and visualization options lie before him. Command and control are at his fingertips.

The manual introduces ALERT II as *a criminal justice information system.* The cover highlights the system's core features, which include information data, operational assistance data, management assistance data, and research and planning data. This is ALERT II's essence. Bits upon bytes upon more megabytes of data.

This *Regional Criminal Justice Information System is designed and operated by the Kansas City, Missouri Police Department to achieve the goals of information availability and operational assistance.*

Figure 16.1. Kansas City, Missouri, Police Department ALERT II Criminal Justice Information System's manual cover.

Before continuing, the manual's fourteen-page introduction gives credit for building the system where it is due. The President's Crime Commission was the system's progenitor, and the manual's introduction quotes extensively from the commission's report.

ALERT II is software. Three separate index files compose its core. The name index file contains key data that unlocks the system's full functionality—an alphabetical list of names, with any and every person ever entered into the system. This includes given names, monikers, aliases, and associated business names.

The general purpose index file uses numbers to identify records stored within the system. These numbers link to an individual's name, property, or vehicle, for example. They can also link to a law enforcement agency record, such as a court case number or a criminal offense code for a crime, such as murder or assault.

For ALERT II, a name provided entrée into a world of connected data. In addition to one's name(s), ALERT II included eight different ways to identify any individual within its system. You could find someone by searching a license or vehicle identification number, an address, a warrant, a criminal record, court case, offense record, prosecutor, or correctional booking number.

The master data file provided a complete record of all information stored in ALERT II pertaining to a given person or vehicle. These could be individual people or profiles. For example, one could retrieve profiles of all people with the family name Johnson, or automobiles manufactured in 1965.

ALERT II was more than just database software. It was a system, and users could input, query, and receive outputs based on the massive amount of information contained in its files.

Like the magazine article said, ALERT II was a cop's best friend. But what was that friend made of? What was its psyche, motivations, and capabilities?

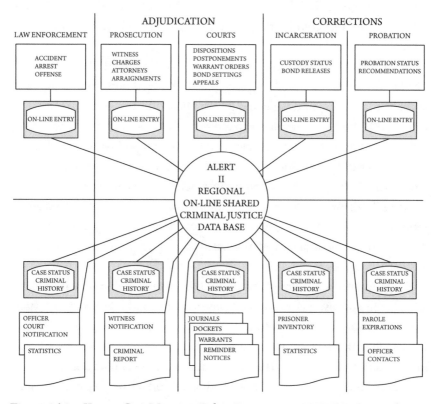

Figure 16.2. Kansas City, Missouri, Police Department ALERT II Criminal Justice Information System's ALERT II database structure.

You don't have to be a psychologist to gain such insights. You just have to understand how its programmers designed it—and why. You have to understand their priorities. You have to know what limits or controls they put in place.

First and foremost, ALERT II's designers prized connectivity. They required a system that could provide real-time data services for all Kansas City law enforcement agencies simultaneously.

They also needed the system to be networked. It had to be able to seamlessly connect, talk to, and exchange information with other state, regional, and national criminal justice information systems. Most important, it had to be able to interface with the FBI's National Crime Information Center (NCIC). But it also had to communicate in an emergency situation. The system needed to be able to transmit an all-points bulletin (APB) through *a* message-switching system that could reach every node in the network. For this, and other reasons, the system had to be built for speed. Whether an APB or system queries from the field, police officials needed designers to build software that could supply requested information through the system within ten seconds.

More for the system's integrity than operability, ALERT II's designers needed the system secured. In principle, authorized personnel only input authentic, accurate, and validated information into the system. ALERT II came with built-in security capabilities. But the Kansas City Police Department and affiliated agencies had to develop and comply with additional, strict protocols, such as where to locate system hardware, where to store backup tapes, and most importantl, decide who had access.

ALERT II's manual doubled as an internal marketing tool. Computer systems like ALERT II were new to the police force, and not necessarily welcomed. So its operator's manual needed to communicate the whys of the system as much as the hows.

Programmers represented the ALERT II system as they saw it. It was a wheel with spokes; and database, teleprocessing, and telecommunications hardware and software powered the system. But law enforcement officials interacted with the system using one or more of eleven peripheral spokes— or application subsystems. These individual software applications provided specific tools while authorized users had access to each of the eleven applications. Users tended to use the application for a specific purpose; law enforcement personnel used law enforcement applications, court personnel used court applications, and so on and so forth. All of these separate applications were integral to the system as a whole. And any one of them had the power to grind—rightfully or not—a citizen in its bureaucratic gears.

But the law enforcement subsystem is what the police command used to find its technological fix. It was their solution to the social, political, and criminal threat posed by black people in Kansas City.

The ALERT II law enforcement application was an information environment all unto itself. Using it—strategically or inadvertently—police could document, profile, surveil, track, and target criminal suspects using race as a key factor. Like ALERT II as a whole, the application consisted of a number of sub applications. The *arrest* system produced data about individuals arrested for crimes. This included statistical information about their criminal arrest history. The *dispatch* system produced data that police command used to allocate human resources and deploy equipment. It also produced statistical analyses of these resource allocation and deployments. The *offense* system documented criminal offenses. And it produced statistical and historical analyses of crimes entered into the system.

When a Kansas City police officer made an arrest, he was required to fill out a paper report. The arrest report alone required the officer to input some thirty-eight distinct data points, including the offender's name and address, race, sex, and age. They also entered detailed physical descriptions like height, weight, and hair and eye color. It even asked for the name of the person's current employer.

A flow chart walked ALERT II's users through the process from arrest to the monthly and annual generation of statistical reports.

When an officer made an arrest, he or she had to submit an offense report, called Form P.D. 189. If the offense involved a victim, then the officer initiated a related report, Form P.D. 339—a *crimes against persons report*. A crime involving a vehicle required yet another report. The offense report required data about the named offense itself. It also required an array of data about the circumstances surrounding the offense. What time did it happen? Was a weapon or force used? Were fingerprints or other evidence collected? Who was the victim? Who reported the crime?

The most important information required from the form, however, was information about the suspect(s). The form provided space for information on two suspected perpetrators. If officers identified a perpetrator(s) they specified the suspect's name, race, and sex, his or her age, physical attributes, and more. When completed, the officer passed the form on to a supervisor. The supervisor then passed it on to the Data Control Unit. They processed it according to uniform protocols. The offense record application was programmed to produce—like clockwork—monthly and cumulative annual reports on all offenses.

The address at which an arrest is made is required and during the on-line entry of arrest information the data is passed through a census tract and block lookup. This is performed by loading address information into a key and reading an on-line file containing the census tract and block corresponding to the address.

This data was then incorporated into the arrest record. Finally the full arrest record was written onto the name, general purpose, and master index file tapes.

The Kansas City Police Department relied heavily on these reports. Police personnel transferred the summary tapes that recorded these offenses to the FBI. The FBI then input the data into the NCIC database.

At the end of each month a program is run that reads the general index and master files and creates a monthly arrest tape which is used as input to the arrest report programs. The monthly tape provides a comprehensive list of statistics on the person arrested, the particulars of the arrest, disposition information, previous arrest record information, and location information.

Finally, the dispatch system tracked and collected data about every officer dispatched to the field to respond to an incident. The reports tracked dispatch location. Most important, however, they tracked how long it took for the officer to arrive on the scene after dispatched. These data were collected and processed in the same manner as the offense records.

* * *

From a law enforcement perspective, ALERT II was a success. It increasingly allowed the Kansas City Police Department to remotely wage a war on crime. This distancing was not insignificant, especially for police patrolling the beat. Echoes of the 1968 uprisings still reverberated across the nation into the 1970s.

In 1973, President Richard Nixon tapped Chief Kelley to head the FBI. Part of Kelley's job was to rehabilitate the agency's image. By that time, the nation had begun to grasp fully the significance of COINTELPRO, the government's secret spying program. The FBI had conspired with army intelligence officers to surveil and create new computerized databanks on anyone the bureau felt "threatened" the national order, including civil rights activists like Martin Luther King Jr., members of the Black Panther Party, and student activists.

As Kelley moved up, the new chief, Joseph McNamara, moved in. The KCMOPD continued to be fully vested in its efforts to mobilize criminal justice information systems. That year—1973—Kansas City law enforcement personnel made more than eight million inquiries into the ALERT II system.[1] Name, case number, and dispatch inquiries topped the list. The Computer Systems Division was located in the department's Administrative Unit. They added a systems analyst technician to the team, and two senior analyst technicians. Those selected for the positions commanded a salary equivalent to the police rank of major. A programmer technician and a supervisory technician also joined the team that year. Computer systems and data processing were seen as part of legitimate police work. In his first year's report as chief, McNamara extolled ALERT II's virtues.

Computerized telecommunications systems have resulted in a 400 percent increase in productive information available to the officer patrolling streets. Mobile terminals in police cars represent the next step forward in providing the officer out in the field with immediate access to criminal data and continued protection of the citizen's privacy. . . . Field officers now have every police record of criminal activity available instantly in the front seat of their patrol cars.

But this was just the tip of the iceberg. Between 1968 and 1973, police officials had begun to understand and experiment with the full capabilities that ALERT II provided.

Kent Colton, an MIT researcher at the time, explained that law enforcement exploited three primary criminal justice information system capabilities. Some law enforcement personnel viewed the system as a pure information carrier. Others saw the potential for transforming and manipulating the data to serve a broader purpose. Still others realized the power of modeling. Like those that Simulmatics had pioneered, these law enforcement officials learned to build computational models. Those models represented crime in either the present—profiling. Or they built frameworks to model the future—forecasting.[2]

When ALERT II debuted in 1968, police personnel used it mostly to replicate mundane data collection and documentation tasks. It tracked traffic tickets, routine court appearances, and the like. But in subsequent years, the department input massive amounts of data into the system. That's when police began to produce, and regularly review, statistical reports on policing

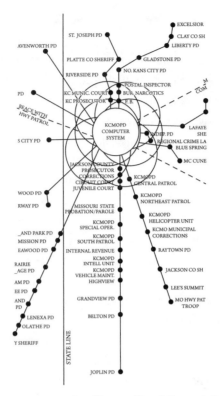

Figure 16.3. Agencies connected to Kansas City, Missouri, Police Department ALERT II Criminal Justice Information System's network.

activities. By 1973, the Kansas City Police Department was ready to move into the third phase—profiling, modeling, and forecasting.

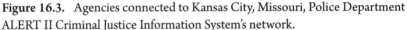

* * *

ALERT II had been operational for six years. Yet, crime increased in Kansas City in 1973. More alarming, the number of robberies—a federal index crime and a part I crime in the KCMOPD nomenclature—had increased by 11 percent over the previous year. McNamara had determined that the problem warranted the full use of ALERT II's powers. That meant it was time to include mapping, profiling, and forecasting into a tactical strategy designed to seek and destroy Kansas City's robbery problem. It had to determine when, where, and how robbery is perpetrated, and by whom.

That year, the department launched *operation robbery control*. Police brass set up a special command and control station. It operated twenty-four hours a day, seven days a week. The unit strategically occupied the 28th Street corner. Command personnel and crime analysts staffed the unit.

Computer data depicting high-frequency time periods and geographic areas particularly vulnerable to armed and strong arm robberies was utilized to deploy the available manpower. Computer analysis, as well as spot maps indicating the location and type of robbery throughout the city were maintained and each robbery plotted as soon as it occurred.[3]

As planning turned to tactical operations, officers in the field saturated patrol areas. They strategically placed decoys. Police disguised themselves as everything from cabbies—for mobile surveillance—to elderly women—to lure purse-snatchers.

What led the special unit to determine where these robberies were most likely to take place? How did they know what number of officers they should deploy in the targeted area? How did they know what type of suspect they should be on the lookout for? How did they know how long it would take for officers to respond if and when they drew out the criminals they set out to neutralize?

Saul Gass had provided the algorithm years before; ALERT II provided the necessary hardware and software to execute it. Start with the police beat. Add census blocks indicators. Add race. Add sex. Add age. Add weighted crime data, a prior criminal offense and arrest history. Then utilize IBM's inferential statistics software like SPSS to analyze, model, and forecast.

These were the key data to be plugged into Gass's algorithm and the tools to process them. McNamara marshaled it all for *operation robbery control.* ALERT II's routine data input, analysis, and output capabilities provided the individualized data and statistics used to deploy it. When you look at what it produced—the maps, tables, and statistical analyses—you could write the correct end to Operation Robbery Control's story, with surprising accuracy.

But McNamara said it all. *The dedication and commitment of the officers assigned to this special and unique project began to pay gratifying dividends. Instead of the customary increase, the rate of robbery by Christmas time had been reduced by almost 27 percent, compared to the same period of 1972.*

* * *

Approaching the late 1960s, Roy Wilkins had grasped the complex, yet still developing relationship between computing and America's so-called "race problem." He did so perhaps more than any other civil rights leader. Wilkins was born of Mississippi stock. His grandfather *had been a slave one minute, a*

ARRESTS BY SEX AND RACE

Part I Offenses	Total	White Male	Colored Male	Other Male	White Female	Colored Female	Other Female
Murder & non-negligent manslaughter	115	19	87		1	8	
Manslaughter by negligence	4	2	2				
Rape	182	60	117		5		
Robbery	951	148	737	2	13	51	
Assault-aggravated	645	202	339	2	28	74	
Assault-not aggravated	3,042	1,357	1,275	2	162	246	
Burglary	2,339	958	1,241	2	76	62	
Larceny	3,534	1,015	1,391	4	435	683	6
Auto Theft	890	352	501	2	18	17	
Total Part I	11,702	4,113	5,690	14	738	1,141	6
Part II Offenses							
Arson	44	20	22		2		
Forgery & counterfeiting	145	43	50		25	27	
Fraud	493	169	170	4	62	88	
Embezzlement	55	25	20		5	5	
Stolen property	91	36	43		10	2	
Vandalism	708	384	238		39	47	
Weapons	885	283	527		22	51	2
Prostitution	945	215	123	1	243	363	
Sex offenses	384	288	74		19	3	
Narcotic laws	1,342	669	530		97	45	1
Gambling	1,663	102	1,526	1		34	
Offenses vs. fam. & child	360	137	210		9	4	
Driving while intoxicated	5,323	3,716	1,227	24	290	66	
Liquor laws	739	368	273	1	52	44	1
Drunkenness	2,711	1,821	637	39	158	51	5
Disorderly conduct	4,915	2,255	1,708	8	474	467	3
Vagrancy	108	44	47	2	5	10	
All other off. exc. trfc.	11,311	4,884	3,955	14	1,626	829	3
Total Part II	32,222	15,459	11,380	94	3,138	2,136	15
Careless driving	10,098	6,071	2,222	69	1,319	415	2
Speeding	29,025	19,191	3,615	104	5,077	946	92
Other traffic violations	55,035	30,254	12,275	54	9,497	2,943	12
Total Other Arrests	94,158	55,516	18,112	227	15,893	4,304	106
GRAND TOTAL	138,082	75,088	35,182	335	19,769	7,581	127

Figure 16.4. Arrests by sex and race, Kansas City, Missouri, 1973.

freedman the next, Wilkins wrote in his autobiography. He also noted that his grandfather's emancipation came with the caveat that *he was still black all day long.*[4]

Reconstruction in 1865 brought his grandfather hope, and the vote, Roy recalled, only to have the Ku Klux Klan and the Mississippi Black Codes steal them both back by the end of the decade. But Grandfather Wilkins bequeathed audacity to his son Willie, who got an education and married Mayfield Edmundson. Edmundson's "café au lait complexion" made Roy rightfully suspicious; his maternal roots found their way back to its seminal master.[5]

Roy knew one thing for certain. His father, Willie, had earned a reputation as a "bad nigger" in Holly Springs, Mississippi, though he once beat a white farmer for addressing him as such. Journalist Ida B. Wells once explained "Lynch Law."

If a colored man resented the imposition of a white man and the two came to blows, the colored man had to die, either at the hands of the white man then and

OFFENSES BY DIVISION AREAS

Part I Offenses	Northeast Division	South Division	Central Division	Not Stated	Total
Murder	14	21	44	2	81
Manslaughter	15	18	9		42
Rape (Force)	45	65	88	23	221
Rape (Attempt)	23	35	23		81
Robbery (Armed)	287	555	610	16	1468
Robbery (Strong Arm)	168	303	385	9	865
Assault-agg. (Gun)	240	188	278	17	723
Assault-agg. (Knife)	106	93	234	12	445
Assault-agg. (Other weapons)	208	170	260	9	647
Assault-agg. (Hands)	50	35	54	6	145
Assaults, Other	473	277	349	24	1123
Burglary (Residence)	2082	3393	1752	94	7321
Burglary (Non-Residence)	827	1044	1147	55	3073
Larceny Over $50	1730	2591	1868	185	6374
Larceny Under $50	1920	2582	2321	195	7018
Auto theft	1033	1367	1461	23	3884
Total Part I	**9221**	**12737**	**10883**	**670**	**33511**

Part II Offenses					
Arson	49	56	39	1	145
Forgery & Counterfeiting	119	141	150	7	417
Fraud	243	309	281	12	845
Embezzlement	94	79	105	6	284
Vandalism	1742	1868	1027	110	4747
Sex offenses	105	142	52	9	308
Kidnapping	10	10	10		30
Casualty	203	231	363	14	811
Dead body	278	245	392	11	926
Animal bite	815	821	363	31	2030
Sick call	1	3	2	2	8
Suicide	11	20	14		45
Attempt suicide	137	132	128	6	403
Lost property	134	215	253	1329	1931
All others	215	253	278	49	795
Total Part II	**4156**	**4525**	**3457**	**1587**	**13725**
GRAND TOTAL PART I & II	**13377**	**17262**	**14340**	**2257**	**47236**

Figure 16.5. Offenses by division areas, Kansas City, Missouri, 1973. Central Division includes highest concentration of black residents.

there or later at the hands of a mob that speedily gathered. If he showed a spirit of courageous manhood he was hanged for his pains, and the killing was justified by the declaration that he was a "saucy nigger."[6]

Willie Wilkins was a saucy nigger. But he hopped aboard the Illinois Central Railroad line with his wife and escaped Mississippi *one-step ahead of a lynch rope.*[7]

By the time Roy Wilkins was born in 1901, his mother and father had settled in St. Louis, Missouri. There they had witnessed something completely foreign. White people mixed with blacks. At train stations, "white" and "colored" signs did not adorn bathroom entrances. On the trolleys, black folks sat where they damn well pleased. White people showed his parents kindness. Some Negroes owned property, in some neighborhoods that were integrated. And Negro physicians—like the one who delivered Roy into the world—were not an anomaly.

But even this northern bastion had its limits. There was a color line. And, after a while, Roy explained that his father had begun to lose the will to even

Part I Offenses	Northeast Division	South Division	Central Division	Not Stated	Total
Murder & non-negligent Manslaughter	24	15	76		115
Manslaughter by negligence	1	2	1		4
Rape	42	38	102		182
Robbery	198	299	452	2	951
Assault-aggravated	221	122	298	4	645
Assault-not aggravated	908	644	1,479	11	3,042
Burglary	810	656	859	14	2,339
Larceny	1,027	1,190	1,295	22	3,534
Auto Theft	274	277	338	1	890
Total Part I	3,505	3,243	4,900	54	11,702
Part II Offenses					
Arson	17	9	18		44
Forgery & Counterfeiting	43	25	76	1	145
Fraud	130	163	197	3	493
Embezzlement	25	7	23		55
Stolen property	42	17	30	2	91
Vandalism	207	187	306	8	708
Weapons	253	180	442	10	885
Prostitution	53	93	793	6	945
Sex offenses	65	71	245	3	384
Narcotic laws	399	449	482	12	1,342
Gambling	456	173	1,019	15	1,663
Offenses vs. family & children	80	74	203	3	360
Driving while intoxicated	877	761	995	2,690	5,323
Liquor laws	259	171	304	5	739
Drunkenness	469	334	1,895	13	2,711
Disorderly conduct	1,591	1,121	2,171	32	4,915
Vagrancy	39	37	32		108
All other off. except traffic	2,705	2,793	5,719	94	11,311
Total Part II	7,710	6,665	14,950	2,897	32,222
Careless driving	2,264	2,770	2,047	3,017	10,098
Speeding	1,427	2,015	347	25,236	29,025
Other traffic violations	6,156	6,701	9,457	32,721	55,035
Traffic traffic arrests	9,847	11,486	11,851	60,974	94,158
GRAND TOTAL	21,062	21,394	31,701	63,925	138,082

Figure 16.6. Arrests by division areas, Kansas City, Missouri, 1973. Central Division includes highest concentration of black residents.

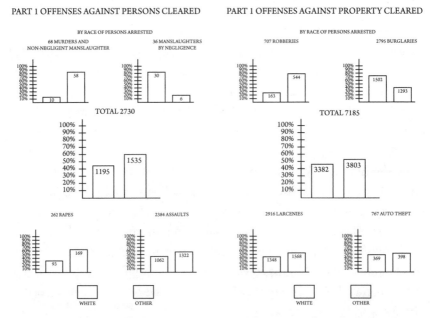

Figure 16.7. Part 1 offenses against persons and property cleared, Kansas City, Missouri, 1973.

think about crossing it. When Roy's mother died, Roy was about six years old. His aunt convinced his father to let Roy and his two siblings go live with her family in St. Paul, Minnesota.

It was in St. Paul that Roy Wilkins witnessed black prosperity and all its possibilities. His uncle had become Howard Elliott's private rail car porter. Elliott was a Harvard-trained engineer, who later became president of the Northern Pacific Railroad. Roy's uncle held a position that earned him not just respect, it also provided a middle-class lifestyle. Roy glimpsed a life where race seemed not to matter.

Perhaps I'm a sentimentalist. But no one can tell me that it is impossible for white people and black people to live next door to one another, to get along—even to love one another. For me integration is not an abstraction constructed on dusty eighteenth century notions of democracy. I believe in it not only because it is right, but because I have lived it all my life.[8]

Following his uncle's advice, Wilkins chose to attend the best high school in St. Paul, George Weitbreit Mechanical Arts High. He had decided to pursue a career in engineering. But a teacher recognized his writing ability, and urged him to develop it. *With that my plans to become an engineer melted in a new lust for books and writing.*[9]

Soon Wilkins discovered W. E. B. Du Bois's *Souls of Black Folk*. His mind grasped DuBois' central concepts: double consciousness and the color line. But, he admitted that at the time, *I was young, Jim Crow had left me mostly alone, and while I loved the beauty of DuBois's writings, his arguments were still a little abstract for me.*[10]

That changed in the summer of 1920. That's when Wilkins said *I lost my innocence on race once and for all.* A police dragnet in Duluth, Minnesota, had ensnared ten Negroes suspected of raping a white woman. During the middle of the night, an angry white mob stormed the jail. They dragged six of the men out to the central square. There, a kangaroo court pronounced three of them guilty. They lynched each one, one by one, from a street light pole, while five thousand whites cheered.

Wilkins worked for a decade as a practicing journalist. Then, he went to work for the organization that began to wage war on the kinds of racial injustices that Wilkins had witnessed for himself. Wilkins began his career as the NAACP's assistant secretary. During that time he became the editor of Du Bois's *The Crisis*, the NAACP's official magazine. In 1955, Wilkins became the NAACP's executive secretary. And in 1964 he was its executive director.

Perhaps Roy Wilkins's early capacity for engineering enabled and motivated him to deeply contemplate the computer revolution's meaning for black people. Perhaps only someone who had witnessed firsthand both racial violence and hope could simultaneously transpose a deep sense of both optimism and pessimism onto the computer. If nothing else, certainly, his journalist's education and longtime editorial work are what allowed him to so adroitly frame the deeply probative question printed in the headline of his September 11, 1967, *Los Angeles Times* editorial. There Wilkins conjectured, *Computerize the Race Problem?*[11]

* * *

Weighing a ton, and standing five feet, one inch at the shoulders, Hattie was a *big-boned, broadfaced, black beast with four white stockings and a white-splashed rump.*[12] Actually, it was black and white. Its color is what would have drawn Wilkins's attention. So would the fact that it was assigned ID number 6069325 in a computerized databank known as a "herd book." He would have noted that it was genetically designed with the aid of a computer and would have paid special attention to the fact that it had to be fed massive amounts of inputs (data/food) to produce value. And he would have paid keen attention to the automated system that monitored and provided daily feedback about her milk production.

This is the kind of Holstein-Friesian milk cow that set Wilkins's mind *working on facets of our ever irritating race question.* The Holstein-Friesian was a crossbreed. Its genetic mixture produced black-and-white–colored cattle equipped for superior milk production. The fact that a 1967 computer could harmonize black and white into a superior milk cow, and automate the production of something with supreme value captivated Wilkins's imagination. What might the computer do if we let it loose to process America's most pressing national problem?

Regardless, the computer performed significant work for Wilkins. The computer was a problem-oriented business machine. Wilkins wondered whether the computer could solve America's race problem; engineer racial equality. If that were too lofty a goal, Wilkins at least questioned how the computer might change black people's relationship to work, career advancement, and economic success.

To Wilkins, the computer was the great displacer.

In Michigan computers are bringing thousands of different cars off the assembly line equipped by orders on punch cards. In Ohio a machine drills 250 cylinder blocks at once. In Louisiana a computer fills drums with chemicals. In Chicago a computer mixes and bakes cakes. On a single Dixie plantation a cotton-picking machine displaces 50 Negro families.

From one perspective, Wilkins's computer was race neutral, merely performing the task it was designed to do. From another vantage point, computers were doing work traditionally done by Negroes. As far as Wilkins was concerned, both led to the unemployment line.

The predicament, Wilkins acknowledged, was both personal and structural. In one breath, Wilkins would admonish the next generation to get an education. He warned that if they did not, *computers may be tossing most of the race to the slag heap by 1990.*[13]

In the next breath he'd say something like, *statistics don't mean much to people who think and live by slogans, but the fact is that three million Americans are hunting for work and cannot get a job. In the midst of our affluence we simply have not solved the problem of what to do about workers dislodged by technology, not by laziness. One element of the formula for unemployment is a dark skin. It is true nationally that the unemployment rate for Negroes is twice that for whites, but in some localities it is as much as four times the white rate. In the Watts district of Los Angeles last August one out of every three Negroes was unemployed.*

For Wilkins, the computer could have also been the great equalizer, were Negroes both attuned to and provided more than token access to the machine. Wilkins once recalled a story for his *New York Amsterdam News* readers. A black student protest broke out at Brandeis. Wilkins harshly criticized the student protestors. He claimed that young folk those days expected to run the joint, before they'd even taken part in the education it had to offer them. Nevertheless, the students had taken over the campus communications center to stage their protest. The center housed a switchboard and a computer.

One of the students had shoved a list of demands in the face of Brandeis's administration. They were—as he put it—non-negotiable. But Wilkins explained, *Well as the young man will learn, most things in this world are negotiable. Instead of spouting this nonsense and wasting precious time he and his fellows should be studying and trying to understand the computer in the Center.*

Wilkins described the irony behind the black student protest.

At about the same day they were throwing their blackness around the Brandeis campus, Dr. Vincent Harding, president of a black college in Atlanta, was bemoaning the drain of black brains from black schools. While he cries for more and better Negro graduates, black students are frittering away their time and the race's destiny by capturing computers before they learn basic algebra.

Nestled between criticisms lay Wilkins's truth. The computer was a ticket to both higher education and upward economic mobility.

But Wilkins's computer was also the great regulator; the great enforcer. Even in its infancy, Wilkins's computer had come to represent the white power structure. The Man. Our national slipstick. Our persistent prejudices associated with blackness marked each one.

New York Citizens were treated recently to what has become the standard riot photograph: black looters loaded with foodstuffs stolen from a Brooklyn super-market. Thus Wilkins opened his *Los Angeles Times* editorial. He was making a point about the magnitude of the image compared to how relatively trivial the infraction was.

He went on to drive a simple, seemingly not apparent, argument home. *Stealing is not racial; the* idea *planted in the public mind that Negroes steal and white people do not is known to be false by every policeman and by every prosecutor in the United States.*

He also raised the point that Alexander's Inc. (a then-popular, discount department store) increased by 20 percent the number of shoplifters it apprehended that year. As a result, Wilkins highlighted, *computers are being used and a list of 250,000 confessed or convicted shoplifters is being offered stores by the security office of one chain.* Wilkins was referring to a *New York Times* story about Spartan Industries, Inc. It was a large retail chain that boasted *the only national file of this kind and the only one using computerized facilities.*[14] In addition to computerizing a national retail and grocery store shoplifting databank, the company also pioneered the use of Knogo Wafers. The anti-theft tags affixed to retail goods were developed alongside related CCTV and other electronic surveillance systems to deter shoplifters.[15]

In any case, the story made Wilkins fearful about how the computer would be used. He knew that white America associated black people with crime. He was afraid that that association, and data that confirmed it, would be fed into, ingested in, and processed by a powerful new computer system—one that stored, connected, and distributed large amounts of decision-driving data that could negatively impact black people's lives.

You see, Wilkins's computer was also cold, impersonal, and unfeeling. Like New York's Board of Examiners, Wilkins argued, the computer *screens applicants, gives examinations, promulgates lists and feeds the transmission belt which cannot pause for unorthodoxy.*[16]

* * *

Despite all of these ways that the computer oriented his gaze toward race, blackness, and black people, it is critical to know that Wilkins's computer was not merely a metaphor. It was a sign. Speaking about more than just the Negro farmworker, Wilkins said that *the computer is but one more signal that he has been kept at arm's length while the rest of America pressed forward into the computer era. In the mass he never got a chance to acquire the learning and the skills which would have enabled him to progress toward the use of data processing.*

As these words suggest, Wilkins's computer mediated racial struggle. In this epic affray, the computer stood somewhere between black people's intractable strivings to gain access to opportunity, and white America's stubborn and belligerent refusal to grant it. In Wilkins's imagination, granting access to opportunities afforded by mastering the computer would be tantamount to inciting some kind of technological insurrection. But the persistent experience of rising up and marching forward, only to be beaten back down, Wilkins argued, weighed heavy on black America's collective psyche.

When I look now into the skeptical eyes of bewildered white observers of this riotous summer, I think of the Arkansas sharecroppers, of Anthony Crawford, a Negro farmer who was lynched because he was "too prosperous." ... The psychological children of this treatment are the 1967 rioters.

Wilkins hoped to train his computer's processing power on this computational color line—not by marshaling a rhetorical imperative or by running a computer command. He simply framed an intricate, three-dimensional question. The question challenged America and the masters of its political machinery. It challenged the designers of its burgeoning technological systems, the puppet masters of private enterprise—and their most prized new digital tool.

Wilkins asked the question in a language each of them had come to understand—the language of data processing. Nudging aside his sacred Holstein, Wilkins asked, *After the computer has defined, on tape, the ideal Holstein, could it then turn its impersonal, unprejudiced magic upon our agonizing race problem? Could it not, after digesting the facts which whites and blacks have fogged over for so long give us an outline of our obligation? Instead of being a measure of the Negro's lag, cannot the computer become a guidepost to interracial justice and peace?*

Wilkins had hoped that the dawn of computing would train a watchful eye on America's race problem. It did. Just not in the way he imagined when his mind wondered about the new technology's possibilities. ALERT II existed. It was used to its full capabilities, and was deemed successful in Kansas City. Special projects like Operation Robbery Control continued. And the Kansas City Police Department added new hardware and software to its expanding arsenal. Through it all, East Kansas City felt increasingly over-policed and under constant surveillance. But this isn't the whole story.

By 1975, ALERT II had sprawled into an expansive network of 227 terminals, utilized by 51 agencies. It extended throughout much of Missouri, and over the state line into Kansas.[17] The computer center housing ALERT II's hub hardware, software, and technical expertise entertained hundreds of law enforcement visitors. They came from across Missouri and Kansas. Scores more rushed in from Canada to Texas, Washington, DC, to Iowa, California to Japan, to Saudi Arabia, and Venezuela.[18] A third of the computer division's personnel time was spent designing new systems and software.

The system itself had amassed a bank of nearly three million individual records.[19] One quarter of ALERT II's reporting function was spent producing statistical offense and arrest reports (among the eighteen different categories of reports). All told, the KCMOPD had spent more than $2 million on the ALERT II's technical systems alone. And this did not include personnel and related costs.

ALERT II became a national model for criminal justice information systems. In fact, the first annual survey of operating criminal justice systems across the United States took place in 1970. The voluntary survey determined that 488 such systems had been built, at least one in every US state, territory, and district.[20] That number continued to skyrocket. Almost two-thirds of the systems were built using IBM hardware, and many were designed and built by IBM contractors.

ALERT II had its own Kansas City story. But the new values, presumptions, and worldview that shifted as a result of its use permeated law enforcement thinking throughout the United States. ALERT II—and systems like it—exemplified black people's relationship to computer technology. The contours of this relationship were wire framed by the late 1960s. They were fully structured and sedimented by the mid-1970s. What was the computer to black people? Black people, by and large, did not have access to the technology being used to profile, target, and forecast their tendency toward criminality. Black people were not hired as technicians to process the data input in the machine. Black people certainly did not design the systems. Black people were not at the table to contribute to conversations about how to deploy the outputs. Black people were not represented among the industry consultants who showcased the computer's capabilities, developed its use cases, or had the technical skills to know how to build in necessary constraints. Black people were scarce in the ranks of students at higher-education institutions that provided the pipeline to government agencies and industries that were so invested in this work.

What was the computer to black people throughout the 1960s and early 1970s? They certainly never saw themselves pictured in an IBM advertisement utilizing one of the new machines for all kinds of business purposes. To black people by and large the computer was an alien technology destined to go to work on them the way all prior US technologies had—to grind them into submission and exert racial power over their entire existence.

And what were black people to the 1960s and 1970s computer system? ALERT II and the rise of criminal justice information systems more broadly began the long-standing process of turning black people—people with bodies and experiences, hopes and interests, aspirations and legitimate grievances against their government—into abstract data. Computers process and manipulate data. Whether they anticipated it or not (and all evidence points to not), once law enforcement officials invested in computing as a law enforcement, problem-solving tool, they also invested in the value of data as an enterprise. The "data" produced the model, and the model became the territory. Crime became something committed by people with whom one has no relationship—certainly not one rooted in mutuality, common experience, common goods or interests, shared responsibility, or, in many cases, even basic humanity.

But it is easier to occupy and place under almost complete and perpetual surveillance an entire community when objective and unbiased "data" tell you this is a prudent course of action. And it is much easier to wage a war against an 11.5 percent robbery rate than it is to have to take into account real people while one tries to make a community safer. It is much easier to believe one's harassment is righteous when an algorithm and software sift through criminal histories and tell you that the source of your crime problem is the whole of East Kansas City.

Chapter Seventeen

DIGITAL TECHNOLOGY

OUR PAST IS PROLOGUE

The year 2000. That's when the dotcom bubble burst, along with many of the Vanguard's dreams. The once-mighty giant America Online had introduced the world to the Internet. Now it had to merge with Time Warner to stay competitive. Both David and Malcolm had already cashed out. They made a fortune and took it with them. But they left *NetNoir* and the Black Internet behind. They had accomplished something truly great and consequential for black America: to think that blackness was at the center of the Internet universe, something responsible for ushering the masses online, even if it was only for a short time. But NetNoir's business model couldn't sustain itself. As the Internet changed phase, David made clear where his interests lay.

We all are a collective body. Let's all move forward, both on a global sense, macro sense, but also within these virtual communities so black folks would just be able to finally improve our lives. That was certainly part of my thing. But I was definitely an entrepreneur. I am an entrepreneur. An entrepreneur in the sense that I intended to have a successful business and then get a return. I wanted to become wealthy enough to become comfortable to live the life I wanted to lead.

Barry followed close behind. AOL bought *Black Voices* in 2004. In part, AOL wanted to replace what it lost with *NetNoir*. AOL wedded Time

Warner, and then *Huffington Post*. New personalities ran in and out of the site as it changed hands. Through it all, Barry remained proudest that its name never changed. *Black Voices is a powerful brand that will last over time.*

GoAfro had died a quick death back in 1998. Nothing and no one took its place. The magazine to which it was tethered also folded soon after. MetroServe stuck it out for another ten years. But GoAfro's dissolution dashed William's hopes. He would never parlay it into a fruitful online venture.

Ken had worked tirelessly, trying to evangelize black Boston to get online. But he could not sell enough software to make a living. So he had to go work for other people, but not particularly his own. His BlackFacts.com remained his primary online legacy.

Kamal never even tried to survive the ruptured bubble.

AfroLink came around, and it was disruptive. There's been nothing around since that's been disruptive.

Kamal decided that disruption was no longer possible. At least, it did not seem plausible. Not for him. Not for AfroLink. Not for any others of the Vanguard. He closed shop and turned to his art.

Some of the Vanguard failed, some succeeded. Others didn't see their lives in such black-and-white terms. Tyronne Foy's BBS didn't last long once the big ISPs came to town. He looked back and realized he had spent a lot of money, had made very little, and said he wouldn't have changed a thing. Lee Bailey's EURweb.com would be a lone survivor through the future. Farai Chideya continued to tell stories about race, politics, or anything she wanted, really.

The Universal Black Pages had been dying a slow death since 1996. But Derrick buried it completely in 2002. He didn't lament its passing. Perhaps the best hope was to take IT to the streets so it could survive the bubble—to engineer change and foment a revolution that would significantly undermine, destabilize, even crumble America's racial order. But in 2006 he lost a friend who had made this her agenda. On September 20, 2006, the *Washington Post* published her obituary.

Anita Brown, 63; Pushed Internet Use in Black
Community
By Yvonne Shinhoster Lamb
Washington Post Staff Writer

Wednesday, September 20, 2006

Anita Brown, 63, who in the late 1990s overcame her fear of technology and became a major proponent of the Internet in the black community, died on Sept. 8 of cardiac arrest at Washington Hospital Center. She lived in the District.

In 1996, Mrs. Brown, once an avowed technophobe, founded Black Geeks Online to promote computer literacy and educate others about the power of information technology. She sought to bridge what came to be known as the "digital divide" by creating her virtual information clearinghouse. Her grassroots effort in Washington flourished with a monthly newsletter and her recurrent Heads-UP e-mail bulletins. For five years, she kept her growing network informed on new media developments, software and hardware changes, company announcements and job openings.

Early on, she pushed to connect African Americans in Washington and elsewhere to the Internet through workshops and seminars. One of her first efforts in 1997, "Taking IT to the Streets," attracted more than 200 people and featured a live feed to chat rooms on the Internet. In 2000, the Boston Globe *said Mrs. Brown was "one of those individuals that author Malcolm Gladwell describes as a 'connector.' Brown has established herself as an unofficial Internet griot who sends out missives to members all over the world, from the South Bronx to South Africa," the article said. "The organization is a digital grapevine for people of color online, and Brown had made it her mission to encourage black information tech professionals to volunteer in the community."*

Mrs. Brown was known as "a serial entrepreneur" by family members for her many business start-ups, including a desktop publishing business. Her most popular business is It's a DC Thang T-Shirts, which she started in the late 1980s. Her interest in expanding her T-shirt business to other cities led her to the Internet. At first she was distrustful of the Web. "I thought it was Big Brother," she once said. "We didn't have any business on it."

But after one of her media-savvy brothers changed her mind, she became hooked. "She really was obsessed with it," said her sister Janet Dyson of Upper Marlboro. "You could find her on the Internet anytime of day or night."

In 1996, she designed and managed the help desk for NetNoir on AOL, which focuses on African American culture and lifestyles. She simplified the technical jargon, added her photo and served as "SistahGeek" concierge. She later moderated a weekly spiritual forum for NetNoir.

Civil rights leaders in the early 1960s had done what they could to arrest the computer revolution's negative impact on black America. They didn't stand much of a chance. By the end of that decade it became abundantly clear: the Negro—America's greatest problem—would be the new computer society's first major problem to solve. Government, industry, and higher education institutions collaborated, designed, built, and deployed automated policing systems, networked databases, and algorithmically driven predictive policing imperatives. Six hundred criminal justice information systems were being used by police, courts, corrections, and other criminal justice agencies up through 1980. Those tools began to lock us up at skyrocketing and racially disparate rates.[1]

But we began to get a hold of those computing tools. We began to master them. We recognized that we could use them to make our individual lives better, more enjoyable, more purposeful. We discovered we could use them to connect ourselves, bind ourselves together, build community, produce culture, create markets, make money. Maybe even start another civil rights revolution. But all that possibility seemed to fade with the Vanguard's collective passing by the middle of the first decade of the new millennium.

* * *

On February 26, 2012, George Zimmerman gunned down Trayvon Martin in Sanford, Florida. Zimmerman was white. He claimed that Trayvon—a black seventeen-year-old—walked suspiciously through his gated community. Sanford police briefly detained Zimmerman then let him go home.

One week passed. Then two. Still no one knew Trayvon's name.

Then, it started.

March 8. Change.org, an online petition site, pressed police to arrest Zimmerman. Two million people signed it.

March 9. The FBI and the Department of Justice launched an investigation.

March 16. Sanford, Florida, police released the 911 tapes from the night Zimmerman shot and killed Martin. The same day, the *New York Times* published African American columnist Charles Blow's *The Curious Case of Trayvon Martin*.

March 19. The Internet and activists staged the Million Hoodie March to protest Zimmerman's freedom and racial profiling.

March 23. Barack Obama, America's first black president, stood before the world at the White House. From news streaming sites, such as YouTube, tethered to fiber-optic cables that stretch across the world, he said:

When I think about this boy, I think about my own kids. My main message is to the parents of Trayvon Martin. You know, if I had a son, he'd look like Trayvon.[2]

April 11. Police arrested Zimmerman for Martin's murder.

The Web, Facebook, Twitter—these and many other platforms carried the raging and virulent news to every corner of the digitally connected world. It was a victory for a new concept we called digital activism.

* * *

The Internet of 2012 was a different world than it was when the Vanguard first made its mark. It had been more than a decade since the 2000 bubble burst. Blackness, black people, black content, black culture, black interests— we were visible in 1994 and 1996 and 1998 and 2000 in a way that it seemed we could never be in 2012. The black footprint on the Internet universe was relatively enormous in the 1990s. Roughly three million websites composed that universe. The dominant Internet portals like AOL and CompuServe shared a stake in black visibility. They had heavily invested financial resources to produce and distribute black content throughout every corner of that still-small universe.

By 2012 that universe had grown exponentially. More than a half billion sites dotted the landscape. And the provincial portals that once invested heavily in steering users to black content suddenly had little stake in doing so. Those walled gardens came down. The Web opened. Every site competed for visibility. And visibility started to become as prime a commodity online as it had always been off, particularly for building wealth, influencing public policy and political interests, and building a movement.

Predictably, as online visibility increased in value, black people's visibility within the open Web began to diminish. The more the Web expanded, the more black people, black content, and black interests receded from view. Search engines like Google became the dominant traffic cop, steering users one way or another until the Internet looked as segregated as any American city, and the Black Internet was a ghetto in all of them; a place where black people, black culture, black content, and black community could scarcely be found and had little value.

This did not mean of course that the Web's black population diminished as content users, producers, or distributors. It just meant that we had to find new ways to connect, direct attention to ourselves, remind people of our worth and value, call attention to the fact that we were still here, and point out that our fates were still connected. Others needed to know that too many of us were still suffering under the weight of an American social, economic, and political structure that sees us as different, inferior, and dangerous. In many ways we needed another Watts, a national wake-up call, a moment to focus all eyes and ears on our collective voice, a voice that reminded everyone that *the evils which can be endured with patience as long as they are inevitable, seem intolerable as soon as hope can be entertained of escaping them.*

The Vanguard, the 1990s Black Internet—they had given us hope. But the policing, the police brutality, the incarceration, the wholesale devaluation of black life were becoming simply intolerable. Fortunately, we found not only our voice, but new online platforms that allowed us not only to showcase who we are, what we produce, and the interests that concern us. We seized new online platforms capable of amplifying our voices on our own terms.

* * *

Technology writer Jenna Wortham told a story for the *New York Times Magazine.*

On July 2013, a 32-year-old writer named Alicia Garza was sipping bourbon in an Oakland bar, eyes on the television screen as the news came through: George Zimmerman had been acquitted by a Florida jury in the killing of Trayvon Martin, an African-American teenager. As the decision sank in, Garza logged onto Facebook and wrote, "Black people. I love you. I love us. Our lives matter." Garza's friend Patrisse Cullors wrote back, closing her post with the hashtag "#blacklivesmatter."[3]

The hashtag had disappeared before it even really spoke, or at least before anyone heard it back in 2013. But the hashtag didn't really matter, according to racial justice activists like Charlene Caruthers. Caruthers is the founding national director of Black Youth Project 100 (BYP100). The youth organization focuses on dismantling structures of anti-blackness through organizing and mobilizing direct action. BYP100 was itself once a hashtag, the moniker for the meeting that helped to conceive a movement still yet to be born.

So the same night that Alicia penned black lives matter as an idea, was the same night that we were gathered. At the time I didn't actually know Alicia, Opal, or Patrisse. I actually met them later through training led by a black organizer. We were in the six-month leadership development training together and that's when I met all three of them.

Before the meeting where she met Alicia Garza, Opal Tometi, and Patrisse Cullors, Charlene had attended the "Beyond November Convening." The convening took place just outside of Chicago, Illinois. University of Chicago political science professor Cathy Cohen had organized the convening. Its purpose was to recruit young activists and train young organizers, particularly those representing the nation's black queer and trans communities. This, according to Charlene, is where BYP100 was born.

We were introduced to the world through a video. We wrote a statement and we put the statement on video and we posted on YouTube and within a day or two it had over 20,000 views. That led to us forming an identity as like an actual entity. The video was released right after our convening. It was released that week. It was our statement to Trayvon Martin's family and to the black community at large. That's how people learned about us through social media and we began to form an identity in the wake of people being curious about "Hey here's a group of young black folks who are articulating what they think is happening, how they feel, and what they want black people to do."

Charlene seemed like she was channeling the late Anita Brown when she emphasized time and again that *people* build movements. Not technology. Not the Internet. Not social media. Like Anita, Charlene recognized communication technology's power, even its necessity. Both recognized that connecting and organizing people was the foundation to one day mobilizing a revolutionary movement. But what happened on August 9, 2014, and the months following made us question yet again. What relationship do black people have with digital media technology? Could we really use it to revolutionize America's racial order? Or at least influence it in that direction?

* * *

I JUST SAW SOMEONE DIE OMFG @TheePharaoh tweeted at 1:03 in the afternoon of August 9, 2014.

I'm about to hyperventilate.

@allovevie the police just shot someone dead in front of my crib yo.

Fuckfuck fuck.

The last tweet featured a photograph. A lifeless black body lying bloodied and still on a Ferguson, Missouri, street. A white police officer stood over his body. His hand still hovered atop his gun.

Retweeted and circulated some seven thousand times, these words and images broke the news to the world. A white, Ferguson, Missouri, police officer had shot and killed a black, unarmed teenager.

By late afternoon, a small crowd had gathered. The boy's body still lay stretched out in the street. There was no ambulance; no coroner there. A local TV station whose faint attempt to report the news via Twitter went virtually unnoticed. As the afternoon turned to evening, everything seemed to stand still. The world had no idea what just happened. And why would they? The housing project where Brown was killed was just another black ghetto. Black people. White cops. Young black male lying dead in the street. Victim of a trigger-happy cop who saw danger lurking in every black body. Was this really news?

The life less body of the seventeen year old kid please help us expose this attempted coverup.

At approximately 7:42, Twitter user @Tefpoe, aka War Machine III, tweeted this message along with a photograph that gave a closer look. A young, black male lay face-down, head to the side, on the pavement. As one looked toward his feet, you could see his shorts resting near his ankles, underwear exposed. From the other end, a stream of blood had gushed from his head and traveled down the empty street. The white cop stood behind him, so as not to have to look into his empty eyes. More than five thousand other users retweeted @TefPoe's call and the image he posted.

Still, little news traveled beyond Ferguson. By mid-evening, bits began trickling out. At 7:31 P.M., the online black news site *NewsOne* ran the headline *Ferguson, Missouri Crowd After Fatal Shooting of Unarmed Teen: "Kill The Police."* At 9:11 P.M., *ABC News* online posted, *Police Shooting in MO Sparks Angry Protests*. One minute later, Raw Story carried, *Officer-related shooting of teenager in St. Louis draws protestors, massive police response*. Then at 9:16 P.M., a breakthrough from Global Grind: *HE HAS A NAME: Ferguson Police Fatally Shoot Unarmed Teenager Michael Brown (PHOTOS)*.

Other sites—Buzzfeed, Mediaite, the *LA Times*, the *New York Daily News*, CBS News, the Blaze, Huffington Post—followed, posting reports before midnight the day Mike Brown's shooting occurred. Mind you, none of these reports that went out in the first twenty-four hours featured eyewitness reporting by journalists on the ground in Ferguson. All of them sourced their information, in whole or in part, from Twitter. In a rare moment, Ferguson's black community got to share with the world what had happened to Mike Brown—what routinely happens in their community. Their voices were unvarnished, unfiltered. And they established a frame for their story that resonated with black America: the great-grandmothers whose sons, husbands, and daughters hung from southern police lynch ropes during Jim Crow; families who watched police kill their loved ones in Watts, in Harlem, in Detroit, in Newark, in Washington, DC, in Kansas City throughout the civil rights uprisings of the 1960s; the young father who scrawled words on a cardboard box and held it up for all of Ferguson, all of Twitter, all of those attuned to online news on that August 9 night: *Ferguson police just executed my unarmed son.*[4]

The *white cop shoots unarmed black teenager* frame that Ferguson set in the first hours following Mike Brown's killing set the stage for what soon erupted.

* * *

August 10, 2014, was an absolutely incredible scene. The world witnessed a gun battle, and a clash of wills in the middle of the streets of Ferguson, Missouri; a large army of policemen, donning Kevlar helmets, most of them in full riot gear, carrying automatic weapons, flash-bang stun grenades, and rubber bullets, and brandishing tear gas. National Guardsmen, riding AR-15 Mega rifle–mounted armored tanks and Humvees; bodies of hundreds and thousands of blacks standing defiantly, with their hands up—asking law enforcement not to shoot, but almost daring them to shoot; acres and acres of broken glass; burned, looted stores and offices. And now this hunt. As police tracked activists, journalists, and citizens calling attention to what was happening in Ferguson, like something out of a bad war movie.

The scene beginning May 10 in Ferguson, Missouri, resembled the uprisings in Watts almost fifty year before. Protestors began to flood Ferguson streets. Then came the police, the National Guard, the national news cameras, the eyes of the world. Watts was purely a military exercise. Ferguson was turning into a true battleground, a literal and virtual world stage where black America began to confront America's racially terroristic legal and law enforcement system head-on, without once blinking.

* * *

People like Allen Frimpong, a young activist and community organizer, helped to pull initial efforts together.

We were talking about going down to Ferguson to be in solidarity with the protests that were happening there. At that point in time we realized we didn't just want to go down there and be in solidarity. We needed to do some relationship building if we're going to go. I chose not to go on that ride to Ferguson. At the time the ride was just to be a ride with friends. That ride started off with just us going became a ride that was national, that had more than one thousand people nationally that went on that ride and that ride from August 28 to September 1 of 2014.

It was during that time—Patrice Collors was one of the founders, creators of the #blacklivesmatter hashtag—we got on the phone to coordinate what became a national ride to Ferguson. It was organized on Facebook. In some of the cities we had relationships…people from the city signed up. And Darnell created a YouTube video that was an invitation for people to join the ride. There was a WordPress site they got created. Soon after friends from other cities began to organize their own rides. We soon got the crowdfunding online that people could raise their own money to go on those rides. I then coordinated with Darnell, Patrice, and others to make sure the core group of folks who were organizing were able to meet with organizations for black struggle in New York and St. Louis the week before so that that relationship could be built and actually engage in activities and capacity building and support for local organizations.

The self-organizing of the WordPress site and the crowdfunding source is what people did in their own communities in order to get to Ferguson. When people got to Ferguson—a week before a core group of people nationally that was orchestrating the national ride—[they] met with local organizers in Ferguson. And when they met with local organizers in Ferguson it was to assess the conditions of what was happening. What resources were needed.

So, for example, for the New York ride we didn't just ask for anybody to go on the ride.…So if you are black you could go on the ride. If you are not black, there were specific instructions about things that you could do to still support the work in Ferguson whether that was to donate, whether to write a letter, or whether to organize a teach-in in your local community.

By the end of August, most of the police and National Guard had left Ferguson. News media trucks lowered their satellite transmitters and headed back to their network bases. Activists who had descended on Ferguson from across the country returned to their homes. Local Ferguson community activists and citizens maintained the pressure after they left. Everyone wanted #justice for #mikebrown. Part of that justice was to indict and bring to trial the man who shot and killed him.

* * *

Black Lives Matter. Up through the initial Ferguson protests it remained just a hashtag—a rarely used one. In the days since Mike Brown was killed, hashtags like #Ferguson, #MikeBrown, #JusticeForMikeBrown, and #Handsupdontshoot circulated exponentially more widely than the hashtag that Garza, Cullors, and Tometi had collectively coined in the wake of George Zimmerman's acquittal for Trayvon Martin's murder. That all changed on November 24, 2014. That was the day that St. Louis County, Missouri, prosecutor Robert McCullough announced that the grand jury refused to file charges against Mike Brown's killer.

On November 24 alone, Twitter users posted 3,420,934 tweets responding to the verdict. Ferguson protestors flooded back into the city streets. The police, National Guard, news cameras and journalists, out-of-town activists and organizers all followed.

Then, it was as if the grand jury's decision not to indict triggered a collective memory for those in, heading to, and watching far away from Ferguson; for those who had demanded justice and had received none. It was like we searched our collective memory for the right words; a phrase that could express what we felt about America's criminal justice system; a phrase we felt represented us as actors with influence and power, not as victims; one that would build us up and help to tear the system down; one perfectly tailored to circulate to the ends of the world through Twitter, other social media platforms, the open Web, and beyond' one that would bind us together and rally us to fight. We reached back and found #BlackLivesMatter.

From November 24, #blacklivesmatter marked not just a moment, but a movement.

* * *

When I went to Ferguson, there were many people you could meet with who were doing many actions at once.

Dream Hampton is a filmmaker, a writer, a storyteller, an activist, with a history that she brought with her to Ferguson.

Patrisse was one of the first people I heard talk about this, as Black Lives Matter being also an extension of the Black Liberation movement. That was very useful for me in seeing it as a continuum of a tradition. Her using those monikers in that way helped me understand my role. What I would say is that as an early adopter to Twitter, repurposing it for social justice. So if you were to go on my site, dreamhampton.com, you'd get information about demanding #JusticeFor-RenishaMcBride. So early on, I thought that there was a need to lift up the names of some of the women who are typically erased, both as victims and as organizers, in this movement, historically. Again really with Aiyana Jones who was a seven-year-old who was killed here in Detroit.

Years ago, I stopped writing. I was one of those people who wrote because my posts were $3.50 a word. When I wrote for Harper's Bazaar—*that was my last piece—I only did one article for them...I was receiving $2.50 a word. When those rates fell off a cliff, I stopped writing. I wasn't someone who woke up every day and needed to write. I never took a journalism class. I was an NYU film student who happened to wander into this magazine called* The Source *in 1990 when they were still a forty-page black-and-white. So I just began writing early on. I was taken under the kind of wings of really great editors like Robert Christgau, and learned how to be a real writer.*

All I ever wanted to do was make films, in terms of professionally. When writing stopped paying, I stopped writing. So around the same time, social media was taking off. I got on Facebook first like everyone, and then to Twitter. I was really late to Facebook and really early onto Twitter (because of other journalist friends). My current Twitter page says [I came on] in 2010, but really I think I came on in 2009 or 2008. When I got too many followers, I killed my account. I never wanted to have more than six figures or more than 50K followers. I'm not someone who thinks that visibility is a great thing, and I found it unmanageable.

Anyway, there are two things happening. I came to New York in 1990 and I come from a very political city, Detroit. I come from that city from the 1970s and 1980s where we had black political leadership and capital. I was political, and began at twelve years old organizing against apartheid. Come to New York and need to get organizing against police brutality. We formed the New York chapter of the Malcolm X Grassroots Movement [MXGM]...which has done two noteworthy things in relation to this movement. We are responsible for the lawsuit against the City of New York for Stop and Frisk. David Floyd who filed the suit against NYC is one of our members, so we spearheaded that suit.

The second thing was that we built a report called "Every 36 Hours," which was counting, in the absence of any federal data collection, counting the number of black people—not just men—who have been extra-judicially killed since Trayvon. Which was then changed to "Every 28 Hours," which then gained traction after Mike Brown was killed. That organization was one that I was a member of for fifteen years.

During those fifteen years, I also accepted a charge from one of my mentors, Assata Shakur, who is of course is on the same Most Wanted List now as Bin Laden....And she asked us to use hip-hop to raise awareness about political prisoners in the United States...[like] Mumia Abu-Jamal who had large profiles. We did hip-hop concerts for ten years called "Black August." So people like Mos Def, Erykah Badu, Fat Joe...we took delegations down to Cuba to perform. I did that for ten years.

In MXGM, we also did a program for many years called Cop Watch. So we would use cameras that I would check out from the NYU Film Department and follow Giuliani's people around until two in the morning as they rode up on kids and then we would give kids "know your rights" pamphlets. That's how I got involved.

When you look at people like Patrisse, using Assata's words from her book as a call...not only to action, but a close-out to all of the gatherings, her thing about "we have nothing to lose but chains" I was the comms director of Hands Off Assata Campaign. When you look at this issue of police terror, it gained traction.

I was one of those early people, along with many other people, but MXGM in particular in NYC, who have been ringing the bell around this since at least Amadou Diallo.

Part of that seismic shift speaks to the relationship that black people and people of color have with computer and digital media technology today.

* * *

The television studio was perfectly chilled that morning of December 4, 2014. The sting of winter escorted me from my Brooklyn home to WNET's midtown Manhattan location. I knew why I was there. I was an "expert"; a contributor; a talking head among others talking heads. But I did not know who the other guests would be. I composed myself while I awaited our moment under the lights, sipped some water, and scanned the news on my phone.

Dr. Butts is running a little late a producer soon informed me. The look on her face said there's no way the show starts without him. I had never met the Reverend Dr. Calvin O. Butts III. But I knew enough about the historic Abyssinian Baptist Church he pastored. It was a Harlem Renaissance icon and a civil rights movement institution. Congressman Adam Clayton Powell Jr. once preached and legislated from its pulpit. I was also familiar enough with Dr. Butts's decades-long efforts to rehabilitate the black image in white New Yorkers' collective mind.

Bronx City Councilman Ritchie Torres soon joined us, and our cast was complete. While we did not know one another, we all knew what brought us to that studio. Ferguson's fires still raged. But a Staten Island grand jury had just ripped open an old wound when it decided not to indict NYPD officer Daniel Pantaleo the day prior. Black Lives Matter was becoming our national organizing framework as Eric Garner's dying words *I can't breathe* became our collective wail, expression of solidarity, and cry for justice.

The day's news agenda presented a number of questions to which Pastor Butts, Councilman Torres, and I would respond. Why are black people so angry? Are the media, social media, and young black activists fanning the flames? Where will the country go from here?

As we readied our responses, Dr. Butts and I discussed the moment's significance.

Of course all of this is nothing new.

Dr. Butts said this as he ticked off the names of black men killed, over decades, at the hands of New York City police alone. Patrick Dorismond. Amadou Diallo. Nicholas Heyward Jr. Malcolm Ferguson. Ousmane Zongo. Sean Bell. This was the same Dr. Butts who had once claimed that *every urban rebellion, every riot, whether in Newark, N.J., or Harlem, N.Y., was started because of police violence and police misconduct.*[5]

Who could argue with Dr. Butts's truth? The deaths that brought us together in that public television studio were but the latest reminders. Police violence persistently threatens people of color. Dr. Butts also reminded me that the unrest, protests, riots, and other responses to these killings were also nothing new. Again, how could I argue with history?

So what was significant about Eric Garner's killing, and its aftermath in New York City? Why was Mike Brown's killing, and the protests and police violence that followed, something that should inevitably pique our attention? Was there more to it than the same old, same old: white cop, black victim, move on? Dr. Butts reminded me that this was all nothing new. But relying on not much more than my intuition I replied simply.

Yes. But there is something that is definitely different.

But, Dr. Butts was right. Whether we're talking about black people's relationship with the police, our relationship to technology, or the overlap between the two—can we ever outrun our history?

ENDNOTES

✳

CHAPTER 1

1. Clemson University, *The Tiger* 81, issue 11 (1987): 6, https://tigerprints.clemson.edu/tiger_newspaper/2238.
2. Ibid.
3. *Massachusetts Institute of Technology Bulletin* 100, no. 2 (November 1964): 466.
4. Ibid., 473.
5. The center was jointly operated by MIT and Harvard.
6. Edward C. Banfield and James Q. Wilson, *City Politics* (Cambridge, MA: Harvard University Press, 1967).
7. Robert F. Wagner Jr., review of *City Politics*, by Edward C. Banfield and James Q. Wilson, Harvard University Press and MIT Press, *The Harvard Crimson*, https://www.thecrimson.com/article/1963/11/19/city-politics-pafter-debating-james-wilson/.
8. Nathan Glazer and Daniel P. Moynihan, *Beyond the Melting Pot: The Negroes, Puerto Ricans, Jews, Italians, and Irish of New York City* (Cambridge, MA: MIT Press and Harvard University Press, 1963).
9. James F. Wagner Jr., review of *Beyond the Melting Pot*, by Nathan Glazer and Daniel P. Moynihan, *The Harvard Crimson*, https://www.thecrimson.com/article/1964/4/8/beyond-the-melting-pot-pin-1908/.
10. Simson Garfinkel and Harold Abelson, *Architects of the Information Society: Thirty-five Years of the Laboratory for Computer Science at MIT* (Cambridge, MA: MIT Press, 1999), ix.
11. *Massachusetts Institute of Technology Bulletin* 100, no. 2 (November 1964): 488.
12. Garfinkel and Abelson, *Architects of the Information Society*, x.
13. Ibid.
14. *Massachusetts Institute of Technology Bulletin* 100, no. 2 (November 1964): 424.
15. Ibid.
16. *Massachusetts Institute of Technology Bulletin* 104, no. 3 (December 1968): 532.
17. Ron Barnett, "Clemson Acknowledges Dark Era in Its History with Marker," *Greenville News*, April 12, 2016, https://www.greenvilleonline.com/story/news/local/pickens-county/2016/04/12/clemson-acknowledges-dark-era-its-history-marker/82944086/.

CHAPTER 2

1. "Bell & Howell, 1975 Annual Report," 13.
2. *Hearings before the United States Equal Employment Opportunity Commission, on Discrimination in White Collar Employment* (January 15–18, 1968), 1.
3. Ibid., 1–2.
4. Ibid., 2.
5. Ibid., 9
6. Ibid., 6.
7. Ibid.
8. Thomas J. Watson and Peter Petre, *Father, Son & Co.* (New York: Bantam, 1990), 369–370.
9. "Producing Power Supplies," IBM Archives, https://www-03.ibm.com/ibm/history/exhibits/brooklyn/brooklyn_3.html.
10. Frank Cary, "IBM Management Principles & Practices," 1974, 7–8.
11. IBM 100, *The Networked Businessplace*, https://www.ibm.com/ibm/history/ibm100/us/en/icons/networkbus/.

CHAPTER 3

1. Michael Walzer, "The Young: A Cup of Coffee and a Seat: A Report from the February 1960 Lunch Counter Sit-ins in Durham, North Carolina," *Dissent* (Spring 1960).
2. N.a., "Students Picket Woolworth's Protest Discrimination in South," N.d.
3. Leo Shapiro, "Boston Group to Back Protest," *Daily Boston Globe*, March 6, 1960.
4. Ibid.
5. Merritt Roe Smith, "God Speed the Institute: The Foundational Years, 1861–1894," in *Becoming MIT, Moments of Decision*, ed. David Kaiser (Cambridge, MA: MIT, 2010).
6. Ibid., 31.
7. "Killian on Technology Future; Stratton to Speak on MIT Role," *The Tech: Newspaper of the Undergraduates of the Massachusetts Institute of Technology*, March 25, 1960.
8. "The President's Report," *Massachusetts Institute of Technology for the Academic Year Ending July 1, 1960*, 3.
9. Ibid., 19.
10. Ibid., 10.
11. "'MIT' EPIC," *The Tech*, March 15, 1960, 2.
12. Ibid.
13. Barry B. Roach, "Don Quixote-ism," *The Tech*, March 15, 1960, 2.
14. "11 Woolworth Stores Picketed," *Daily Boston Globe*, March 20, 1960; *Boston Globe*, 18.
15. Richard Neel Sutton, letter to the editor, *The Tech*, March 22, 1960.
16. Jean Pierre Frankenhuis, letter to the editor, *The Tech*, March 22, 1960.
17. Michael Levin, letter to the editor, *The Tech*, March 25, 1960.

18. *NAACP Bulletin*, June 25, 1960.
19. Personal correspondence between Willam Gamson and James Robinson, April 27, 1960.
20. Emphasis in the original.
21. "Belafonte Sings Here Tomorrow to Aid Students in Dixie Protest," *Daily Boston Globe*, April 20, 1960; *The Boston Globe*, 9.
22. Personal correspondence between James R. Robinson and Willam Gamson, May 4, 1960.
23. Personal correspondence between Curtis Gans and Ella Baker, April 12, 1960.
24. Personal correspondence between Student Nonviolent Coordinating Committee and the Emergency Public Integration Committee, June 14, 1960.
25. Joseph Hanlon, "We Want to Be Segregated Says Black Muslims" *The Tech*, November 7, 1962.
26. "James Baldwin to Talk on Negro Problems at Civil Rights Meeting," *The Tech*, October 24, 1962.
27. "MIT-EPIC Pickets Woolworth's Store, Eight Distribute Leaflets Saturday," *The Tech*, March 8, 1960.
28. "Report of the President for the Academic Year 1970–1971," *Massachusetts Institute of Technology Bulletin*.
29. UMANA Academy, "History of the School," https://www.bostonpublicschools.org/Page/2283.
30. "Science Fair Winners," *Boston Globe*, March 28, 1983.

CHAPTER 4

1. Excerpt from Tom Wells and Hugh Wilson, "WKRP in Cincinnati (Main Theme)," performed by Steve Carlisle, MCA Records, 1981.
2. "An Evening with the Legendary Lee Bailey," http://www.blogtalkradio.com/jay-king/2011/06/07/an-evening-with-the-legendary-lee-bailey, June 6, 2011.
3. DECs' PC model was built to compete with IBM, but was not favorably received.
4. Simson Garfinkel and Harold Abelson, *Architects of the Information Society: Thirty-five Years of the Laboratory for Computer Science at MIT* (Cambridge, MA: MIT Press, 1999), 2.
5. Lesley Oelsner, "2 Indicted in Raid on N.Y.U. Center," *New York Times*, July 30, 1970.
6. Ron Schnell, *Artspeak Reference Guide*, 2012, http://artspeak.quogic.com/artspeak.pdf.
7. Margot Lee Shetterly, *Hidden Figures: The American Dream and the Untold Story of the Black Women Mathematicians Who Helped Win the Space Race* (New York: William Morrow, 2017).

CHAPTER 5

1. Anita M. Samuels, "Black Culture, Computerized," *New York Times*, February 28, 1993, F12.

CHAPTER 7

1. "Record Unit 372, Smithsonian Institution, Office of Public Affairs, Clippings, 1968–1991," https://siarchives.si.edu/collections/siris_arc_216939.
2. Merlisa Lawrence Corbett, "The Business of Bulletin Board Systems: A BBS Can Enhance Your Current Business," *Black Enterprise*, November 1995, 47–50.

CHAPTER 8

1. "Texas Remember: Juneteenth," https://www.tsl.texas.gov/ref/abouttx/juneteenth.html.
2. David Plotnikoff, "Two for the Highway Originators of Netnoir Hope to Spread African American Culture on the Internet," *San Jose Mercury News*, May 15, 1995, 1C.
3. Jon Pessah, "Afrocentric Online Service Debuts," *Newsday*, May 21, 1995.
4. "NetNoir Online, the Cybergateway to Afrocentric Culture…," *Business Wire*, May 22, 1995.
5. Kim Cleland, "Crossover Is Key to NetNoir's Biz Plan," *Advertising Age*, May 22, 1995.
6. "Widening the Net," *Marketing Computers*, June 1995.
7. "IBM-Lotus Takeover Bid," CNN Computer Connection, June 10, 1995.
8. Ibid.
9. Excerpt used with permission by Stephanie Han, a former Net Noir employee. Her full profile of Malcolm Casselle is located online at https://www.stephaniehan.com/wp-content/uploads/2017/08/PROFILEMalcolmCasSelle.pdf.
10. Elizabeth Weis, "New Noir Creates Major Black Presence on the Internet," *Associated Press*, June 19, 1995.
11. "Falling through the Net: A Survey of the 'Have Nots' in Rural and Urban America," National Telecommunications and Information Administration, July 1995, https://www.ntia.doc.gov/ntiahome/fallingthru.html.
12. Encyclopedia.com, s.v. "Farai Chideya," https://www.encyclopedia.com/people/literature-and-arts/american-literature-biographies/farai-chideya.
13. Katie Hafner and Matthew Lyon, *Where Wizards Stay Up Late: The Origins of the Internet* (New York: Simon & Schuster, 1998).

CHAPTER 9

1. A. Appadurai, "The Capacity to Aspire: Culture and the Terms of Recognition," in *Culture and Public Action*, ed. V. Rao and M. Walton (Palo Alto, CA: Stanford University Press, 2004), 59–84.
2. *Record of Society of Actuaries* 12, no. 2 (1986).
3. This and all other cocaine ads are taken from "33 Shameless Cocaine Ads Prove the 70s Were a Hell of a Time to Be Alive," http://canyouactually.com/cocaine-ads-from-the-70s/.
4. Teresa Watanbe, "Where Student Drug Abusers Find Aid," *San Jose Mercury News*, June 5, 1985.

5. Robert Reinhold, "Life in High Stress Silicon Valley Takes a Toll," *New York Times,* January 13, 1984.
6. Jill Jonnes, *Hep-cats, Narcs, and Pipe Dreams: A History of America's Romance with Illegal Drugs* (Baltimore: Johns Hopkins University Press, 1996), 372.
7. Ibid.
8. Ibid., 367–371.
9. San Jose Mercury News archive, https://longform.org/archive/publications/san-jose-mercury-news.
10. Barry Cooper, "Bo's a Rarity: Few Pros Reach Two-Sport Stardom," *Seattle Times,* November 13, 1988, C6.
11. Barry Cooper, "Shaq Just Plays It Cool," *Chicago Tribune,* June 21, 1992.
12. Barry Cooper, "Looking for a New Computer? It's in the Mail; Ordering PCs through Catalog Is Popular, but Drawbacks Are Emerging," *Chicago Tribune,* May 7, 1993.

CHAPTER 10

1. "Taking IT to the Streets, *Wired,* February 7, 2000.

CHAPTER 11

1. Martin Kilson, "The First Congressman from Harlem," *New York Times,* November 7, 1971, https://www.nytimes.com/1971/11/07/archives/adam-by-adam-by-adam-clayton-powell-jr-illustrated-260-pp-new-york.html.
2. 2Pac, featuring The Outlawz, "When We Ride," on Tupak Shakur, *All Eyez on Me,* Interscope & Death Row Records, 1996.
3. Kilson, "The First Congressman from Harlem."
4. Thomas Johnson, "A Man of Many Roles," *New York Times,* April 5, 1972.
5. Donald N. Michael, *Cybernation: The Silent Conquest: A Report to the Center or the Study of Democratic Institutions* 7, no. 17 (Washington, DC: Center for the Study of Democratic Institutions, 1962).
6. Ibid., 1–9.

CHAPTER 12

1. Elaine Brown, "The End of Silence." Album. *Seize the Time: The Black Panther Party.* 1969.
2. "Fortune 500: A Database of 50 Years of Fortune's List of America's Largest Corporations," http://archive.fortune.com/magazines/fortune/fortune500_archive/full/1965/.
3. Mike Conway, "A Guest in Our Living Room: The Television Newscaster before the Rise of the Dominant Anchor," *Journal of Broadcasting and Electronic Media* 51, no. 3 (2007): 457.
4. Ibid.
5. Craig Allen, "Discovering Joe Six Pack Content in Television News: The Hidden History of Audience Research, News Consultants, and the Warner Class Model," *Journal of Broadcasting and Electronic Media* 49 (2005): 355–356, esp. 363.

6. Maxwell E. McCombs, "Negro Use of Television and Newspapers for Political Information, 1952–1964," *Journal of Broadcasting and Electronic Media* 12 (1968): 261.
7. "1964: CBS Television," http://www.peabodyawards.com/award-profile/cbs-reports.
8. "The Open Door," *Journal of Broadcasting and Electronic Media* 10 (1966): 189–190.
9. Ibid.
10. "Transcript of the Johnson Address on Voting Rights to Joint Session of Congress," *New York Times,* http://www.nytimes.com/books/98/04/12/specials/johnson-rightsadd.html.
11. See Keith Breckenridge, *Biometric State: The Global Politics of Identification and Surveillance in South Africa, 1850 to the Present* (Cambridge: Cambridge University Press, 2014), 172.
12. Edwin Black, *IBM and the Holocaust: The Strategic Alliance between Nazi Germany and America's Most Powerful Corporation* (New York: Random House, 2001).

CHAPTER 13

1. The conversation is an amalgamation of interactions detailed in correspondence between Henry S. Ruth and R. E. McDonnel on November 29, 1965, and a December 1, 1965, memo from Henry S. Ruth to James Vorenberg, titled "Visit by R.E. McDonnell of IBM."
2. "Command and Control," *Marine Corps Doctrine Publication* 6 (1996): 45–47.
3. Thomas J. Watson and Peter Petre, *Father, Son & Co.: My Life at IBM and Beyond* (New York: Bantam, 2013).
4. Ibid.
5. Ibid., 97.

CHAPTER 14

1. The National Advisory Commission on Civil Disorders, Hearing Schedule, Thursday, November 9 Room 446, Executive Office Building.
2. October 15, 1967. Memorandum to the Commissioners, the National Advisory Commission on Civil Disorders. Subject: Media Conference in Poughkeepsie; Simulmatics Contract.
3. Ithiel de Sola Pool and Alex Bernstein, "The Simulation of Human Behavior—A Primer and Some Possible Applications," *American Behavioral Scientist* 6, no. 9 (1963): 83–85.
4. This and the following italicized material citing the work of the Simulmatics Corporation are taken from "News Media Coverage of the 1967 Urban Riots: A Study Prepared for the National Advisory Commission on Civil Disorders by The Simulmatics Corporation. Final Report, Submitted to: Professor Abram Chayes, Chairman Media Analysis Task Force by Dr. Sol Chaneles, Project Director Simulmatics Corporation, 16 East 41st Street, New York, New York, February 1, 1968."
5. Ibid.

CHAPTER 15

1. Douglas E. Kneeland, "Behind the Violence: Despair and Spring Madness," *New York Times*, April 12, 1968, 20.
2. Philip D. Carter, "White Things Coming Apart at Seams: Whites Still Live in Calm World," *Washington Post*, June 29, 1969, 1.
3. "Police Chiefs in FBI Survey Assail Lack of Public Support," *St. Louis Dispatch*, December 2, 1964.
4. D. J. R. Bruckner, "Midwest Cities React to Year of Crime, Riots," *Los Angeles Times*, September 28, 1967, 16.
5. KCPT, "'68: The Kansas City Race Riots Then and Now: Town Hall," https://www.youtube.com/watch?v=uW2PZrZh9zY.
6. Ibid.
7. Ibid.
8. Ibid.
9. Kent W. Colton, *Police and Computer Technology: Use, Implementation and Impact* (Lexington, MA: Lexington Books, 1978).
10. Ibid., 324.
11. Ibid., 324–325.
12. "IBM 1961 Annual Report," 7.
13. Ibid., 8.
14. Saul I. Gass, "On the Division of Police Districts into Patrol Beats," *Proceedings of the 1968 23rd ACM National Conference* (ACM, 1968), 459–473, esp. 464.
15. Ibid., 461.
16. Ibid., 460.

CHAPTER 16

1. "KCMOPD 1973 Annual Report," 59.
2. Kent W. Colton, *Police and Computer Technology: Use, Implementation and Impact* (Lexington, MA: Lexington Books, 1978).
3. Ibid., 65.
4. Roy Wilkins, with Tom Matthews, *Standing Fast: The Autobiography of Roy Wilkins* (New York: Viking 1982), 10.
5. Ibid., 15.
6. "(1900) Ida B. Wells, 'Lynch Law in America,'" July 13, 2010, http://www.blackpast.org/1900-ida-b-wells-lynch-law-america.
7. Wilkins, *Standing Fast*.
8. Ibid., 30.
9. Ibid., 38.
10. Ibid., 38.
11. Roy Wilkins, "Computerize the Race Problem?" *Los Angeles Times*, September 11, 1967.
12. Paul O'Neil, "How Now World's Greatest Cow?" *Atlantic Monthly*, September 1973.
13. Roy Wilkins, "The Reaction of the Responsible," *New York Amsterdam News*, August 7, 1965.

14. Isadore Barmash, "Spartans to Share List of U.S. Shoplifters," *New York Times,* June 12, 1970.
15. *Billboard Magazine,* January 30, 1973.
16. Roy Wilkins, "Wilkins Speaks: All They Need Is Clean Paper and a Heart," *Baltimore Afro-American,* December 17, 1966.
17. Alert Network, 35150 National Criminal Justice Reference Service (NCJRS), 2.
18. Ibid., 3.
19. Ibid., 12.
20. US Department of Justice, "Law Enforcement Assistance Administration: 1972 Directory of Automated Criminal Justice Information Systems."

CHAPTER 17

1. Lawrence D. Bobo and Victor Thompson, "Racialized Mass Incarceration: Poetry, Prejudice, and Punishment," in *Doing Race: 21 Essays for the 21st Century,* ed. Hazel R. Markus and Paula Moya, 322–355 (New York: Norton, 2010). https://scholar.harvard.edu/files/bobo/files/2010_racialized_mass_incarceration_doing_race.pdf.
2. "Obama: 'If I Had a Son, He Would Look Like Trayvon,'" March 23, 2012, https://www.youtube.com/watch?v=wAPtUfOs7Gs.
3. Jenna Wortham, "Black Tweets Matter: How the Tumultuous, Hilarious, Wide-Ranging Chat Party on Twitter Changed the Face of Activism in America," *Smithsonian Magazine,* September 2016, https://www.smithsonianmag.com/arts-culture/black-tweets-matter-180960117/#v0swKAgVWDlAo3Qu.99.
4. "Ferguson, Missouri Community Furious after Teen Shot Dead by Police," Huffpost, August 11, 2014, https://www.huffingtonpost.com/2014/08/09/ferguson-teen-police-shooting_n_5665305.html.
5. Angela Mosconi, "Rev. Butts: Police 'Lynching' Has to Stop," *New York Post,* March 20, 2000.

SELECTED SOURCE
DOCUMENTATION

★

PERIODICALS

"All Gary Webb's Dark Alliance Stories in the *San Jose Mercury News*." *San Jose Mercury News*. August 1996. https://www.narconews.com/darkalliance/drugs/start.html.

"An Evening with the Legendary Lee Bailey." June 6, 2011. http://www.blogtalkradio.com/jay-king/2011/06/07/an-evening-with-the-legendary-lee-bailey.

Andrews, Paul. "Afrolink Software Inc. First of Its Kind, Afrolink Software Puts Issues on Line." *Seattle Times*. September 24, 1990.

Barmash, Isadore. "Spartans to Share List of U.S. Shoplifters." *New York Times*. June 12, 1970.

"Belafonte Sings Here Tomorrow to Aid Students in Dixie Protest." *Daily Boston Globe*. April 20, 1960, 9.

"Bell & Howell Sharpens Its Product Focus." *BusinessWeek*, no. 2278 (May 1973): 66–67.

"Big Dig Paves Road to Riches for Many." The Associated Press. June 8, 1998.

"Black Lawmakers Win CIA Promise for Drug Probe." Associated Press. September 20, 1996.

Blackpast. "(1900) Ida B. Wells, 'Lynch Law in America.'" July 13, 2010. http://www.blackpast.org/1900-ida-b-wells-lynch-law-america.

"Blue-Collar Training Gets a White-Collar Look." *BusinessWeek*, no. 2187 (July 1971): 76–77.

Bruckner, D. J. R. "Midwest Cities React to Year of Crime, Riots: Detroit, Tense With..." *Los Angeles Times*. September 28, 1967.

"Carson's Launches IBM Audio Response System in Its Store." *Chicago Defender*. December 18, 1965, 35.

Carter, Philip D. "White Things Coming Apart at Seams: Whites Still Live in Calm World. *Washington Post and Times Herald*. June 29, 1969.

Castillo, Elias. "Silicon Valley Cocaine Users Risking Addiction to Heroin: Latest Coke Chaser Persian Drug Becoming 'Very, Very Serious Problem' Here." *San Jose Mercury News*. September 11, 1985.

"Charles of Blame Fly in Kansas City Disorders: Large Area Sealed Off Battalion of Troops Requested." *Christian Science Monitor.* April 12, 1968, B13.

Cleland, Kim. "Crossover Is Key to NetNoir's Biz Plan." *Advertising Age.* May 22, 1995.

The Clemson University Tiger. January 24, 1987.

——. February 20, 1987.

——. October 23, 1987.

——. November 13, 1987.

——. November 11, 1987.

——. January 22, 1988.

——. January 29, 1988.

——. February 10, 1989.

——. March 3, 1989.

Clendinen, Dudley. "Black's Mayoral Bid Brings Change to Boston. Special to the *New York Times.* October 7, 1983.

Cobb, Nathan. "Suddenly, a Boom in Sites Geared toward African Americans." *Boston Globe.* January 2, 1996, 25.

Coleman, Sandy Coleman. "Starting a Small Business: Hard Work, Great Satisfaction; Sidebar William Murrell, Company President." *Boston Globe.* October 17, 1993.

"College Grads Get IBM Jobs through League." *Chicago Daily Defender.* July 31, 1967, 4.

"Computer Aids Kansas City Police Setup: Warnings Relayed." *Christian Science Monitor.* August 26, 1968.

"Computer Helps Blood Banks Here." *Chicago Daily Defender.* November 2, 1970, 10.

Congbalay, Dean. "Drug Users Trade Cocaine for 'Crank.'" *San Jose Mercury News.* September 15, 1986.

Cooke, Russell. 1987. "Group in Boston Calls for Boycott of Goode over Move." *Philadelphia Inquirer.* June 25, 1987.

Correspondence between Jack Richard and Walter Findlator, Network Administrator, Afra-Span Network—Atlanta. *Boardwatch Magazine.* November 1994, 13.

Dent, Lisa. "Innovative Collaborations." *American Visions Magazine.* October 1, 1994.

"DIT Hit with Race Bias Suit." *Atlanta Daily World.* September 19, 1974, 4.

Eaton, Sabrina. "NASA Helps Blacks Reach for Stars." (Cleveland) *Plain Dealer.* September 16, 1994.

"11 Woolworth Stores Picketed." *Daily Boston Globe.* March 20, 1960, 18.

Encyclopedia.com, s.v. "Farai Chideya." https://www.encyclopedia.com/people/literature-and-arts/american-literature-biographies/farai-chideya.

"Essay Winners Are Named." *Calhoun* (South Carolina) *Times.* December 2, 1982, 1.

Fallin, Glen. "Police Computer Debut." *Afro-American.* October 12, 1968.

"Ferguson, Missouri Community Furious after Teen Shot Dead by Police." HuffPost. August 11, 2014. https://www.huffingtonpost.com/2014/08/09/ferguson-teen-police-shooting_n_5665305.html.

Fletcher, Michael A. "Black Caucus Urges Probe of CIA-Contra Drug Charge." *Washington Post.* September 13, 1996.

"Fortune 500: A Database of 50 Years of Fortune's List of America's Largest Corporations, 1965." http://archive.fortune.com/magazines/fortune/fortune500_archive/full/1965/.

"Free Forum on Cocaine Set Tonight." *San Jose Mercury News.* February 20, 1986.

Glass, Andrew J. "White House Brings Together the Black Internet Community." *Seattle Post-Intelligencer*. September 20, 1999.

"GoAfro!" *American Visions* 10, no. 2. April 1995, 40.

"GoAfro." *Anonymous. American Visions*. April/May 1996, 11, 2, 35.

"Here's What's behind the Changes in Cocaine Prices on US Streets since the 1980s." *Business Insider*. Sunday, October 30, 2016.

"Holds City Bias Keeps after Her." *New York Amsterdam News*. December 15, 1951, 5.

"How Negroes Fight Poverty in Watts: Operation Bootstrap Prepares Negroes for Jobs in Ghetto." *Jet*. December 21, 1967.

"IBM Approves $115,000 Study Grant for UNFC." *Atlanta Daily World*. October 22, 1970, 2.

"IBM Employment Means 'Room for Achievement.'" *Afro-American*. October 17, 1964, 1.

"IBM Expanding Research Demands Huge Work Force." *Afro-American*. September 25, 1965, 1.

"IBM Key Punch Programs at State U. Urban Center." *New York Amsterdam News*. January 27, 1968, 22.

"If the Action Is Futuristic, IBM Will Have a Piece of It." *Afro-American*. March 27, 1965, 41.

Intercity. *BBS Magazine*. November 1995, 31.

Jenkins, Timothy L. "A Gathering of Eagles." *American Visions* 9, no. 3. June 1, 1994, ProQuest 4.

Jenkins, Timothy L. "An Interactive Niagara Movement." *American Visions* 9, no. 2 (April/May 1994).

———. "Want Ad for a Revolution." *American Visions* 8, no. 5 (October/November 1993): 4.

"The Job Gap for College Graduates in the '70s." *Business Week* (September 23, 1972): 48.

Johnson, Thomas. "A Man of Many Roles." *New York Times*. April 5, 1972.

"June 10, 1995. IBM-Lotus Takeover Bid." CNN Computer Connection. June 10, 1995.

Kansas City Panthers, Police Fight. *St. Louis Post-Dispatch*. December 6, 1969.

"Kansas City Police Unveil Alert System." *Los Angeles Times*. July 11, 1968, 4.

"Kansas City's Curfew Calm Broken by Fires and Snipers." *Chicago Tribune*. April 12, 1968, 2.

Kiley, David. "Black Surfing." *Brandweek* 38, no. 43 (November 17, 1997): 36.

Kilson, Martin. "The First Congressman from Harlem." *New York Times*. November 7, 1971. https://www.nytimes.com/1971/11/07/archives/adam-by-adam-by-adam-clayton-powell-jr-illustrated-260-pp-new-york.html.

Kneeland, Douglas E. "Behind the Violence: Despair and Spring Madness." Special to *The New York Times*. April 12, 1968.

Lawrence Corbett, Merlisa. "The Business of Bulletin Board Systems." *Black Enterprise*. November 1995.

"Learning for Earning." *Time* 100 (July 31, 1972): 40.

"Letters." *American Visions Magazine*. December 1, 1996.

Lindsey, Robert. "Police Send Up Copters in Fight on Urban Crime." Special to *The New York Times*. December 9, 1970.

"Lucky Day? IBM Breaks Ground, Finds Horseshoe, But…" *Charlotte Observer*. September 21, 1978.

Martins, Gus. "Catalyst for Activism in Roxbury." *Boston Globe*. June 29, 1988.

"MCI Plan to Deliver Messages." *New York Times*. September 23, 1983. http://web. archive.org/web/20030730064537/www.mcimail.com/binaries/prices.pdf.

McKinney, Hampton. "Chicago Urban League Employment and Guidance." *Chicago Daily Defender*. Multiple dates throughout 1967.

Mosconi, Angela. "Rev. Butts: Police 'Lynching' Has to Stop." *New York Post*. March 20, 2000.

"Multiple Stories on IBM's Building New Locations in Charlotte, North Carolina." *Charlotte Observer*. June 29, 1978.

"NCC Couple Join IBM for Summer." *Afro-American*. May 23, 1964.

The Negro Motorist Green Book: An International Travel Guide. 1949 edition.

"NetNoir Online, the Cybergateway to Afrocentric Culture..." *Business Wire*. May 22, 1995.

"Northern Students Tour Sitdown Areas." *The Shaw* (North Carolina) *Journal*. March–April 1960.

"Obama: 'If I Had a Son, He Would Look Like Trayvon.'" https://www.youtube.com/watch?v=wAPtUfOs7Gs.

Oelsner, Lesley. "2 Indicted in Raid on N.Y.U. Center." *New York Times*. July 30, 1970.

O'Neil, Paul. "How Now World's Greatest Cow?" *Atlantic Monthly*. September 1973.

Page, Clarence. "CIA Drug Link Alarming to Blacks." *St. Louis Post-Dispatch*. September 24, 1996.

Pearman, Robert. "Black Crime, Black Victims." *The Nation*. April 21, 1969.

Perry, Charles. "Freebase: A Treacherous Obsession. The Rise of Crack Cocaine and the Fall of Addicts Destroyed by the Drug." May 1, 1980.

"Personal Technology African Internet Society Organizing." *Atlanta Journal-Constitution*. September 22, 1996.

Pessah, Jon. "Afrocentric Online Service Debuts." *Newsday*. May 21, 1995.

Pinkerton, Stewart. "Watts a Year Later." *Wall Street Journal*. August 12, 1966.

Plotnikoff, David. "Two for the Highway: Originators of NetNoir Hope to Spread African American Culture on the Internet. *San Jose Mercury News*. May 15, 1995.

"Police Chiefs in FBI Survey Assail Lack of Public Support." *St. Louis Post-Dispatch*. December 11, 1966.

"Preparing for the Information Age." *Black Enterprise* 26, no. 8 (March 1996): 66.

"Profile of Malcolm CasSell." Excerpt used with permission of Stephanie Han, a former Net Noir employee. https://www.stephaniehan.com/wp-content/uploads/2017/08/PROFILEMalcolmCasSelle.pdf.

Reinhold, Robert. "Life in High Stress Silicon Valley Takes a Toll." *New York Times*. January 13, 1984.

Ribadeneira, Diego. 1989. "In Drug War, Victories Are Small, Fleeting Demand for Product Keeps an Industry Vibrant." *Boston Globe*. March 31, 1989.

Ridinger, R. B. M. "The Universal Black Pages." *Choice* 34 (1997): 80.

Roach, Ronald. "Black Internet: Search Engine Moves Ahead." *Black Issues in Higher Education* 16, no. 4 (April 15, 1999): 55.

Rogers, John G. "A Cop's Best Friend." *Boston Globe*, January 16, 1972, B23.

"Science Fair Winners." *Boston Globe*. March 28, 1983.

"Search, Seizure Victims Sought." *Boston Globe*. November 29, 1989.

Shapiro, Leo. "Boston Group to Back Protest." *Daily Boston Globe*. March 6, 1960.

"Shaw Senior Wins Award and Position with TBM." *Pittsburgh Courier*. June 25, 1966.

"'68: The Kansas City Race Riots Then and Now: Town Hall." https://www.youtube.com/watch?v=uW2PZrZh9zY.

"Software's Hard Sell: Colored People's Time in the Computer Age." *Emerge* 1, no. 7 (May 1990): 13.

"Symposium: Clarence M. Kelley." *Rockville* 22, no. 2 (March 1, 1968): 98–99.

The Tech. February 14, 1962, 5.

———. October 24, 1962, 1.

———. November 7, 1962, 2.

———. December 12, 1962, 1, 5.

———. January 9, 1963.

———. April 7, 1965, 4.

———. March 22, 1966, 1, 2.

———. September 30, 1966. 1.

———. October 4, 1966, 4.

———. October 14, 1966, 11.

———. February 17, 1967, 8.

———. February 6, 1968, 11.

———. April 9, 1968, 1–8.

———. April 30, 1968, 1, 3.

———. May 3, 1968, 3.

———. October 11, 1968, 1–12.

———. November 19, 1968, 1.

———. December 3, 1968, 1–3.

———. February 14, 1969, 4, 12.

———. February 21, 1969, 1, 7.

———. July 3, 1969, 1.

———. December 18, 1969, 9.

———. March 17, 1970, 1.

———. February 4, 1972, 1.

———. August 1, 1975, 6.

"Texas Remembers: Juneteenth." https://www.tsl.texas.gov/ref/abouttx/juneteenth.html.

Thompson, Garland L. "Mapping the Future: Conference Probes Where the Information Superhighway Will Take Blacks." *Black Issues in Higher Education*. October 6, 1994.

Train Dropouts at Two City Centers." *New York Amsterdam News*. March 21, 1964.

"Transcript of the Johnson Address on Voting Rights to Joint Session of Congress." *New York Times*. March 16, 1965. http://www.nytimes.com/books/98/04/12/specials/johnson-rightsadd.html.

United Press International. "FBI to Review Police Training Project. *Arizona Republic*. August 15, 1972.

Van Gorder, Barbara. "Computers and Technology: Is the Whole World Going On-Line?" *American Visions Magazine*. April 1, 1995.

Vaz, Valerie. "On-line Fever." *Essence* 26, no. 5 (September 1995): 38.

Vendel, Christine, and John Shultz. "Rioting in City Takes Five Lives." *Kansas City Times*. April 11, 1968.

Vilkomerson, David. "Students Picket Woolworth's Protest Discrimination in South." March 1, 1960.

Walzer, Michael. "The Young: A Cup of Coffee and a Seat: A Report from the February 1960 Lunch Counter Sit-ins in Durham, North Carolina." *Dissent*. Spring 1960.

Watanbe, Teresa. "Where Student Drug Abusers Find Aid." *San Jose Mercury News*. June 5, 1985.

Weis, Elizabeth. "New Noir Creates Major Black Presence on the Internet." Associated Press. June 19, 1995.

"We're Running Out of Houses: The Great IBM Home Search." *Charlotte Observer*. July 17, 1978.

"Widening the Net." *Marketing Computers*. June 1995.

Wilkins, Roy. "Computerize the Race Problem?" *Los Angeles Times*. September 11, 1967.

——. "The Reaction of the Responsible." *New York Amsterdam News*. August 7, 1965.

——. "Wilkins Speaks: All They Need Is Clean Paper and a Heart." *Baltimore Afro-American*. December 17, 1966.

——, with Tom Matthews. *Standing Fast: The Autobiography of Roy Wilkins*. New York: Viking, 1982.

"Winning Essay Is from St. John's School, 5th Grade." *Calhoun* (South Carolina) *Times*. November 8, 1979.

Wong, Doris Sue. 1987. "Group Says Alleged Rape Case Still Open Despite Probe's End." *Boston Globe*. March 26, 1987.

Wortham, Jenna. "Black Tweets Matter." *Smithsonian Magazine*. September 2016. https://www.smithsonianmag.com/arts-culture/black-tweets-matter-180960117/#v0swKAgVWDlAo3Qu.99.

ORGANIZATION/CORPORATE RECORDS, CORRESPONDENCE, CONFERENCE PROCEEDINGS, AND ANNUAL REPORTS

Bell & Howell. Annual Report. 1975.

Bell & Howell Schools Advertisements. 1972.

BGSA Information Exchange Committee. "Master email contact list." Distributed by Derrick Brown, Georgia Tech Black Graduate Student Association. October 13, 1993.

Brown, Derrick. "BGSI Preliminary Advertising Sales Organization Plan." Memorandum to Derrick Brown from Derrick Brown. December 16, 1996.

——. "BGSI Preliminary Consulting Action Plan." Memorandum to Derrick Brown from Derrick Brown. January 21, 1997.

——. "Developing Electronic Commerce and Direct Marketing Solutions for Product-Based Black Businesses." Memorandum to Derrick Brown from Derrick Brown. January 27, 1997.

——. "KnowledgeBase Strategic Business Plan (take 1)." October 21, 1998.

——. Participation in First Annual Georgia Tech Computer Expo. Memorandum to Metro Atlanta Computer Vendors. March 1, 1998.

——. "Simple Unix Tricks (Impress Your Friends!)." Black Graduate Student Association. November 1, 1995.

——. "Strategic Business Plan." May 2, 1996.

——. "Strategic Business Plan for KnowledgeBase, Inc."

——. "Using Internet Search Engines and Directories." KnowledgeBase, Inc. June 21, 1999.

——. "Wrapup.txt." June 20, 1994.

——, and Mack Jenkins. "How to Build a Computer (Teacher's Edition)." KnowledgeBase, Inc. April 11, 1998.

——, and Lou Macalou. "UNIX Made Easy (16 Commands You Need to Know)." Black Graduate Student Association. October 26, 1994.

"Congressional Black Caucus Foundation Sponsors 24th Annual Legislative Conference, 'Embracing Our Youth for a New Tomorrow.' September 13–17, 1994." News release. September 1, 1994.

"Dictionary of IBM & Computing Terminology." https://www.ibm.com/ibm/history/documents/pdf/glossary.pdf.

"EPIC Action Letter." Summer 1960.

"IBM Revised Programmer Aptitude Test." http://ed-thelen.org/comp-hist/IBM-ProgApti-120-6762-2.html.

"An Introduction to the ELM Mail System." November 17, 1993.

"The Involved Generation—Computing People and the Disadvantaged by David B. Mayer, IBM Systems Development Division, White Plains, New York. Fall Joint Computer Conference, 1969."

"KnowledgeBase Business Plan." June 13, 1994.

"Letter from President Bill Clinton to The Honorable Kweisi Mfume, chairman, Congressional Black Caucus, Regarding Background Paper on Department of Defense Personnel Gathering Information about American Black Civil Rights Leaders."

"Making Web Pages (The Basics) Teacher's Edition." KnowledgeBase, Inc. April 11, 1998.

"Memo regarding Community Project #2 ('Operation Boot-Strap')." September 9, 1965.

"News for Immediate Release, from Harvard EPIC (Emergency Public Integration Committee)." For Further Information H. Pressman. UN4-5395.

"Outcomes of the CBCF California Regional Planning Meeting." Memorandum to Chairman Alan Wheat and Congressional Black Caucus Members Julian Dixon, Walter R. Tucker III, and Maxine Waters from Quentin R. Lawson, executive director. April 21, 1993.

"Producing Power Supplies." IBM Archives. https://www-03.ibm.com/ibm/history/exhibits/brooklyn/brooklyn_3.html.

"Purpose.txt." Distributed by Derrick Brown, Georgia Tech Black Graduate Student Association. September 27, 1993.

"Report on the Conference on Jobs and Job Training." January 25, 1964.

"Scope.txt." Distributed by Derrick Brown. Georgia Tech Black Graduate Student Association. April 25, 2994.

"Statement of Goals of SDS Activity in Civil Rights, May 16–17, 1960," 1–26.

Cary, Frank. "IBM Management Principles & Practices." 1974.

Clay, William L. "Media Organizations Attending the 24th Annual Legislative Conference." Congressional Black Caucus Foundation Twenty-Fourth Annual Legislative Conference Critique. September 20, 1994.

Congressional Black Caucus Foundation's 24th Annual Legislative Conference. "Embracing Our Youth for a New Tomorrow." September 14–18, 1994.

Correspondence from Mr. William A. Gamson to James R. Robinson. May 4, 1960.

Correspondence from Curtis Gans to Ella Baker. April 12, 1960.

Correspondence from Martin Luther King Jr. and Mr. Gordon Thomas. July 3, 1964.

Correspondence from Lawrence W. M. McVoy to Herbert Hill. August 1, 1961.

Correspondence from Omaha Branch NAACP to Roy Wilkins. August 7, 1961.

Correspondence from James Robinson to William Gamson. April 27, 1960.

Correspondence from James Robinson to Harvey Pressman. June 14, 1960.

Correspondence from James R. Robinson to William Gamson. May 4, 1960.

Correspondence from Ira Solomon to Marion Barry. October 10, 1960.

Correspondence from Student Nonviolent Coordinating Committee to the Emergency Public Integration Committee. June 14, 1960.

Crowley, George, and Louise Crowley. "Beyond Automation." *The Monthly Review.* November 1964.

DEC-11-AJPB-D PDP-11 BASIC Programming Manual. Single-User, Paper Tape Software, 1970.

Dozier, Camille. "Quick Tips on Unix Mail." Distributed by Derrick Brown. Georgia Tech Black Graduate Student Association.

Draft Prospectus. Tougaloo Computer Fund. October 23, 1994.

Georgia Tech Black Graduate Students Association. "The Information Committee of the Black Graduate Students Association at Georgia Institute of Technology P * R * O * U * D * L * Y A * N * N * O * U * C * E * S. …"

IBM. "THINK: Our History of Progress, 1890s to 2001." http://www-05.ibm.com/uk/ibm/history/interactive/ibm_history_2.pdf.

IBM 100. "The Networked Businessplace." http://www-304.ibm.com/jct03001c/ibm/history/ibm100/us/en/icons/networkbus/.

International Business Machines Annual Reports, 1950–1989.

Johnson, Howard W. MIT commencement speech. June 7, 1968.

Kansas City, Missouri, Police Department Annual Reports, 1973.

Locker, Mike. "A Proposal for Initiating Discussion and Internal Debate in EPIC," c. 1960.

Massachusetts Institute of Technology Bulletin & President's Reports, 1950–1972.

McNamara, Joseph D. Kansas City, Missouri, Police Department Annual Reports. Report on Impact of Alert II Criminal Justice Information System. 1975.

N.a. The Negro and Automation/Cybernation on the Agenda." Student Nonviolent Coordinating Committee Conference, 1964.

NAACP Bulletin. June 25, 1960.

National Association for the Advancement of Colored People. Telegram to J. A. A. Burnquist. June 16, 1920.

——. "Westcoast Region Annual Report." 1967.

National Society of Black Engineers History. https://www.nsbe.org/About-Us/NSBE-History.aspx#.XJ5_7utKjOQ.

Newsworthies Unlimited. Organization: Operation Bootstrap (a Community Project). November 15, 1965. Local and National Leaders Aid "Operation Bootstrap."

Operation Bootstrap (a Community Project). October 15, 1965. James Farmer Approves Operation Bootstrap.

Pauling, Linus, et al. *The Triple Revolution*. Santa Barbara, CA: The Ad Hoc Committee on the Triple Revolution, 1964.

Peabody Awards. "1964: CBS Television." http://www.peabodyawards.com/award-profile/cbs-reports.

Proceedings of The Public Labor Education Meeting of the Fourth Triennial and Thirty-Seventh Anniversary of the Brotherhood of Steeping Car Porters/AFL-CIO/CLC. Interview with A. Philip Randolph. September 9, 1962.

Program of the Congressional Black Caucus Foundation 26th Annual Legislative Conference. September 11–15, 1996.

Rustin, Bayard. Automation and the Negro.

——. *Fear, Frustration, Backlash: The New Crisis in Civil Rights.* Jewish Labor Committee. 1966.

——. "Humanizing Life in the Megalopolis." The Institute for Religious and Social Studies. November 9, 1965.

——. The Meaning of Birmingham. *Liberation*. June 1963.

Smith, Merritt Roe. "God Speed the Institute: The Foundational Years, 1861–1894." In *Becoming MIT, Moments of Decision*. Edited by David Kaiser. Cambridge, MA: MIT Press, 2010.

Strategic Marketing Plan (DRAFT). BGS Infosystems, Inc. June 26, 1996.

Student Nonviolent Coordinating Committee. "Invitations Sent to the Following Observers." September 7, 1960.

Students for a Democratic Society. Invitation letter to SDS 1964 Annual Convention.

24th Annual Legislative Conference of the Congressional Black Caucus Foundation. "African Americans in the Telecommunications Age." September 14, 1994.

Umana Academy, "History of the School." https://www.bostonpublicschools.org/Page/2283.

Watson, Thomas J., and Peter Petre. *Father, Son & Co.* New York: Bantam Books, 1990.

GOVERNMENT DOCUMENTS

Alert Network, 35150 NCJRS, 2. Manual created by the Kansas City Police Department. c. 1972.

Bay Area Census. City of Palo Alto. 1970–1990 Federal Census Data.

Carter, Harold J. "Report on Deutsche Hollerith Maschinen, G. m. b. H., German Subsidiary of International Business Machines Corporation." December 8, 1943. "Computers Play a Deadly Game: Cops and Robbers," *Think* (May 1971): 28–30.

"The Challenge of Crime in a Free Society: A Report by the President's Commission on Law Enforcement and Administration of Justice." February 1967.

Chaneles, Sol. "News Media Coverage of the 1967 Urban Riots: A Study Prepared for the National Advisory Commission on Civil Disorders by The Simulmatics Corporation." February 1, 1968.

Cook, Gretchen. Entering the World of Multi-media Technology. *Black Enterprise* 26, no. 11 (June 1996): 262.

Distler, A., W. Flury, B. Forman, C. Moran, and M. Sherwood. "Final Report on the 1972 NASIS/LEAA Survey of Automated Criminal Justice Information Systems." September 30, 1972.

"Document 012898." In *Alert II. System Documentation and User Manual.* c. 1972.

"Falling through the Net: A Survey of the 'Have Nots' in Rural and Urban America." US Department of Commerce. July 1995. https://www.ntia.doc.gov/ntiahome/fallingthru.html.

"Federal Data Banks, Computers and the Bill of Rights. Hearings before the Subcommittee on Constitutional Rights of the Committee on the Judiciary." 92nd Cong. 1971.

Fleming, Harold C. "Community Resources for Expanding Equal Employment Opportunity."

Ginzberg, Eli. Excerpts from address at NAACP 54th Annual Convention. July 2, 1963.

"Head-on Collision Course for Civil Rights and Automation." News from US Department of Labor. Week of August 2, 1965.

Hearings before the Select Committee to Study Governmental Operations with Respect to Intelligence Activities of the United States Senate. 94th Cong. 1975.

Hearings before the United States Equal Employment Opportunity Commission on Discrimination in White Collar Employment. January 15–18, 1968.

Helstein, Ralph. Address at NAACP 54th Annual Convention. July 2, 1963.

Institute for Defense Analyses. "Task Force Report: Science and Technology: A Report to the President's Commission on Law Enforcement and Administration of Justice." 1967.

Kansas City Police Department. Introduction to "ALERT II. A Criminal Justice Information System." c. 1972.

Kelling, George L., Tony Pate, Duane Dieckman, and Charles E. Brown. "The Kansas City Preventive Patrol Experiment." The Police Foundation. October 1974.

"Meeting and Conference Schedule, November 9–12." Memorandum, The National Advisory Commission on Civil Disorders. November 6, 1967.

"The National Advisory Panel on Insurance in Riot-Affected Areas Will Hold Open Hearings Wednesday, November 8, and Thursday, November 9." Document #37. The National Advisory Commission on Civil Disorders, Office of Information. November 6, 1967.

National Crime Commission Records. Correspondence from Henry S. Ruth and R. E. McDonnell. November 29, 1965.

1980 Census of Population: Characteristics of the Population. Chapter A. Number of Inhabitants. Part 6. California.

Parry, Robert. "Nicaragua/CIA." Associated Press. April 14, 1986.

Powell, Adam C. Speech delivered to the National Convention of The National Association for the Advancement of Colored People. July 14, 1961. Philadelphia, Pennsylvania.

Priebe, John A., and Others. "Detailed Occupation and Years of School Completed by Age for the Civilian Labor Force by Sex, Race, and Spanish Origin: 1980 Census of Population Supplementary Report." Bureau of the Census. March 1983.

Randolph, A. Philip. "Statement at Labor Dinner of NAACP Fifty-second Annual Convention." July 14, 1961.

"Record Unit 372, Smithsonian Institution, Office of Public Affairs, Clippings, 1965–1991: Collective Overview." https://siarchives.si.edu/collections/siris_arc_216939.

"Report of The National Advisory Commission on Civil Disorders." March 1, 1968.

Samuels, Anita M. "Black Culture, Computerized." *New York Times*. February 28, 1993.

Schlossberg, Howard. "Curiosity Seeker Turns Hobby into Data-base Business." *Marketing News* 28, no. 2 (January 17, 1994): 9.

Simulmatics Corporation. "Revised Final Report: Improving Effectiveness of the Chieu Hoi Program (U), Volume I. Summary—Findings and Recommendations." September 1967.

US Department of Justice, Law Enforcement Assistance Administration. "1972 Directory of Automated Criminal Justice Information Systems."

———. "1980 Directory of Automated Criminal Justice Information Systems."

US Marine Corps. "Command and Control." *Marine Corps Doctrine Publication* 6 (1996): 45–47.

"Violence in the City—An End or a Beginning? A Report by the Governor's Commission on the Los Angeles Riots." December 2, 1965.

"Visit by R. E. McDonnell of IBM." Memorandum to James Vorenberg from Henry S. Ruth. December 1, 1965.

Wirtz, W. Willard. "Youth—Its Freedom and Responsibility: Youth and Employment." Address to Greater Hartford Forum. November 10, 1965.

ACADEMIC ARTICLES AND CONFERENCE PROCEEDINGS

Allen, Craig. "Discovering Joe Six Pack Content in Television News: The Hidden History of Audience Research, News Consultants, and the Warner Class Model." *Journal of Broadcasting and Electronic Media* 49 (2005): 355–356, esp. 363.

Archive of 1970s–1980s Cocaine Advertisements. http://canyouactually.com/cocaine-ads-from-the-70s/.

Banfield, Edward C., and James Q. Wilson. *City Politics*. Cambridge, MA: Harvard University Press, 1967.

Barnouw, Erik. *The Sponsor: Notes on a Modern Potentate*. New York: Oxford University Press, 1978.

Bass, Bernard M., and D. A. Dobbins. "IBM Mark Sense Cards in Prison Classification and Criminological Research." *Journal of Criminal Law and Criminology and Police Science* 47 (1956): 436.

Black, Edwin. *IBM and the Holocaust: The Strategic Alliance between Nazi Germany and America's Most Powerful Corporation*. New York: Random House, 2001.

Breckenridge, Keith. "The Biometric State: The Promise and Peril of Digital Government in the New South Africa." *Journal of Southern African Studies* 31, no. 2 (2005): 267–282.

Byars-Winston, Angela, Nadya Fouad, and Yao Wen. "Race/Ethnicity and Sex in US Occupations, 1970–2010: Implications for Research, Practice, and Policy." *Journal of Vocational Behavior* 87 (2015): 54–70.

Carey, James W. "Variations in Negro/White Television Preferences." *Journal of Broadcasting and Electronic Media* 10, no. 3 (1966): 199–212.

Colton, Kent W. "Police and Computer Technology—The Expectations and the Results" *AFIPS*, 1979, 433.

———, ed. *The Police and Computer Technology: Use, Implementation and Impact*. Cambridge, MA: Department of Urban Studies & Planning, MIT, 1977.

Conway, Mike. "A Guest in Our Living Room: The Television Newscaster before the Rise of the Dominant Anchor." *Journal of Broadcasting and Electronic Media* 51 (2007): 457.

De Graaf, Lawrence B. "The City of Black Angels: Emergence of the Los Angeles Ghetto, 1890–1930." *Pacific Historical Review* 39, no. 3 (1970): 323–352.

Fang, Irving E. "The "Easy Listening Formula." *Journal of Broadcasting and Electronic Media* 11, no. 1 (1966): 63–68.

Garfinkel, Simson, and Harold Abelson. *Architects of the Information Society: Thirty-five Years of the Laboratory for Computer Science at MIT.* Cambridge, MA: MIT Press, 1999.

Gass, Saul I. "On the Division of Police Districts into Patrol Beats." In *Proceedings of the 1968 23rd ACM National Conference,* 459–473. New York: ACM, 1968.

"Illicit Production of Cocaine." *Forensic Science Review* 5 (1993): 95–107.

Jonnes, Jill. *Hep-cats, Narcs, and Pipe Dreams: A History of America's Romance with Illegal Drugs.* Baltimore: Johns Hopkins University Press, 1996, 372.

Lazarus, Harold. "The Corporate Conscience." In *Academy of Management Proceedings,* no. 1, 135–144. Briarcliff Manor, NY: Academy of Management, 1968.

Lyle, Jack, and Walter Wilcox. "Television News—an Interim Report." *Journal of Broadcasting* (1963): 157–166.

McCombs, Maxwell E. "Negro Use of Television and Newspapers for Political Information, 1952–1964." *Journal of Broadcasting and Electronic Media* 12 (1968): 261.

Michael, Donald N. *Cybernation: The Silent Conquest* 7, no. 17. Center for the Study of Democratic Institutions, 1962.

Moulton, Sally Brewster. "Roxbury, Boston, and the Boston SMSA: Socioeconomic Trends 1960–1985." *New England Journal of Public Policy* 4, no. 2 (1988): 39–61.

Moynihan, Daniel P. (Daniel Patrick), and Charles Herbert Backstrom. *Beyond the Melting Pot: The Negroes, Puerto Ricans, Jews, Italians, and Irish of New York City.* Cambridge, MA: MIT Press and Harvard University Press, 1963.

NASA. "NASA Personnel." In *NASA Historical Databook.* 1999.

"The Open Door." *Journal of Broadcasting and Electronic Media* 10 (1966): 189–190.

Pool, Ithiel de Sola, and Alex Bernstein. "The Simulation of Human Behavior—A Primer and Some Possible Applications." *American Behavioral Scientist* 6, no. 9 (1963): 83–85. https://scholar.harvard.edu/files/bobo/files/2010_racialized_mass_incarceration_doing_race.pdf.

"Reinsurance Underwriting Issues." *Record of Society of Actuaries* 12, no. 2 (1986).

Shetterly, Margot Lee. *Hidden Figures: The American Dream and the Untold Story of the Black Women Mathematicians Who Helped Win the Space Race.* New York: William Morrow, 2017.

Winthrop, Henry. "The Sociological and Ideological Assumptions Underlying Cybernation." *American Journal of Economics and Sociology* 25, no. 2 (1966): 113–126.

PERSONAL INTERVIEWS

Al-Mansour, Kamal. February 21, 2017.

Bailey, Lee. December 16, 2015.

Brown, Derrick. November 5, 2015; December 10, 2015; December 31, 2015.

Caruthers, Charlene, with Emily Parker. December 22, 2015.

CasSell, Malcolm. July 20, 2018.

Chideya, Farai. December 17, 2015.

Cooper, Barry. December 17, 2015.

Ellington, David. November 5, 2015; November 13, 2015.

Foy, Tyronne. July 20, 2017.

Frimpong, Allen, with Emily Parker. December 18, 2015.

Granderson, Ken. May 26, 2016; July 17, 2018.

Hampton, Dream, with Emily Parker. December 16, 2015.

Hohlman, Mike. May 7, 2016; May 10, 2016.

Murrell, William. December 29, 2015; April 24, 2016; February 2, 2018; July 27, 2018.

Onwere, Ken. June 22, 2018.

OTHER

Anastasio, Sal. "In Memory of Jacob Schwartz." *Notices of the AMS*. May 2015.

Brown, Elaine. "The End of Silence." Recording, Vinyl LP. *Seize the Time: The Black Panther Party*. 1969.

Donald N. Michael Obituary. *The University Record*, the University of Michigan, Ann Arbor, Michigan. November 20, 2000.

Du Bois, W. E. B. *The Souls of Black Folk* (Chicago: A. C. McClurg & Co., 1903).

Schnell, Ron. *Artspeak Reference Guide*. 2012. http://artspeak.quogic.+com/artspeak.pdf.

2Pac, featuring The Outlawz. "When We Ride." On Tupak Shakur, *All Eyez on Me*. Interscope & Death Row Records. 1996.

INDEX

★

For the benefit of digital users, indexed terms that span two pages (e.g., 52–53) may, on occasion, appear on only one of those pages.